A TEXT BOOK OF

COMMUNICATION SKILLS - I

For

First Semester of Diploma in Engineering

As Per New Revised Syllabus of SBTE Jharkhand, July 2017

With Latest Question Paper

Prof. B. V. PATHAK

Ex. Lecturer in English, Humanities, Industrial Management
and Engineering Economics at Govt. Polytechnic, Pune
and Visiting Lecturer in Government and
Other Engineering Colleges.

N0668

F. Y. Diploma : Semester I - English　　　　　**ISBN 978-93-81237-93-9**

Seventh Edition :　August 2019

© :　Author

Published By :

NIRALI PRAKASHAN

Abhyudaya Pragati, 1312, Shivaji Nagar,
Off J.M. Road, PUNE – 411005
Tel - (020) 25512336/37/39, Fax - (020) 25511379
Email : niralipune@pragationline.com

➢ DISTRIBUTION CENTRES

PUNE

Nirali Prakashan : 119, Budhwar Peth, Jogeshwari Mandir Lane, Pune 411002, Maharashtra
(For orders within Pune)　　Tel : (020) 2445 2044, Mobile : 9657703145
Email : niralilocal@pragationline.com

Nirali Prakashan : S. No. 28/27, Dhayari, Near Asian College Pune 411041
(For orders outside Pune)　　Tel : (020) 24690204 Fax : (020) 24690316; Mobile : 9657703143
Email : bookorder@pragationline.com

MUMBAI

Nirali Prakashan : 385, S.V.P. Road, Rasdhara Co-op. Hsg. Society Ltd.,
Girgaum, Mumbai 400004, Maharashtra; Mobile : 9320129587
Tel : (022) 2385 6339 / 2386 9976, Fax : (022) 2386 9976
Email : niralimumbai@pragationline.com

➢ DISTRIBUTION BRANCHES

JALGAON

Nirali Prakashan : 34, V. V. Golani Market, Navi Peth, Jalgaon 425001, Maharashtra,
Tel : (0257) 222 0395, Mob : 94234 91860; Email : niralijalgaon@pragationline.com

KOLHAPUR

Nirali Prakashan : New Mahadvar Road, Kedar Plaza, 1st Floor Opp. IDBI Bank, Kolhapur 416 012
Maharashtra. Mob : 9850046155; Email : niralikolhapur@pragationline.com

NAGPUR

Nirali Prakashan : Above Maratha Mandir, Shop No. 3, First Floor,
Rani Jhanshi Square, Sitabuldi, Nagpur 440012, Maharashtra
Tel : (0712) 254 7129; Email : niralinagpur@pragationline.com

DELHI

Nirali Prakashan : 4593/15, Basement, Agarwal Lane, Ansari Road, Daryaganj
Near Times of India Building, New Delhi 110002 Mob : 08505972553
Email : niralidelhi@pragationline.com

BENGALURU

Nirali Prakashan : Maitri Ground Floor, Jaya Apartments, No. 99, 6th Cross, 6th Main,
Malleswaram, Bengaluru 560003, Karnataka; Mob : 9449043034
Email: niralibangalore@pragationline.com

Other Branches : Hyderabad, Chennai

niralipune@pragationline.com　　|　www.pragationline.com
Also find us on [f] www.facebook.com/niralibooks

Preface ...

I have great pleasure in bringing out the first edition of this book as per the newly formulated syllabus effective from July, 2017. Great care has been taken to cover the whole syllabus concerning grammar and language parts prescribed in the new syllabus by the Board of Technical Examinations, Jharkhand.

I am very thankful to Shri. Abhijeet S. Nawgekar for his immense help in editing and making all possible improvements in the text. My thanks are also to **Shri. Dineshbhai Furia, Shri. Jignesh Furia, Mr. Malik Shaikh** and the entire team of **Nirali Prakashan** who have printed this book at the right time of students' need.

I will be glad to receive suggestions on my book, both from teachers and students so as to consider them for inclusion in the next edition of this book. I extend my good wishes to all the teachers and students with a genuine hope that this book will be warmly received by them.

Prof. B. V. Pathak

Syllabus ...

Name of Topic	Hours	Marks
PART I : TEXT • Comprehension – Responding to the questions from <u>Text.</u> [Spectrum] • Vocabulary – Understanding Meaning of New Words from Text. • Identifying parts of speech from the Text.	10	18
PART II : APPLICATION OF GRAMMAR • Verbs • Tenses Do as Directed (Active/Passive, Direct/Indirect, Affirmative/Negative/Assertive, Interrogative, Question Tag, Remove Too, Use of Article, Preposition, Conjunctions, Interjections, Punctuation). • Correct the errors from the sentences.	10	18
PART III : PARAGRAPH WRITING • Types of Paragraphs. (Narrative, Descriptive, Technical). • Unseen Passage for Comprehension.	04	08
PART IV : VOCABULARY BUILDING • Synonyms • Antonyms • Homophones • Use of Contextual Words in a given Paragraph.	06	12
PART V : SOFT SKILL DEVELOPMENT • Speaking Skill • Introduction to Group Discussion • Process of Group Discussion • Leadership Skill • Instant Public Speaking	08	16
PART VI : ETIQUETTES AND BODY LANGUAGE • Telephone Etiquettes Listening / Speaking • Problems of Telephonic Conversation • Verbal/Oral Etiquettes • Physical Appearance • Eye Contact/Body Language • Group Discussion	04	08

****

Contents ...

***

Part – I

Comprehension & Vocabulary

(1) Comprehension – Responding to the Questions from **Text (Spectrum)**
(2) Vocabulary – Understanding Meaning of new words from Text.
(3) Identifying Parts of Speech.

[Marks 18]

COMPREHENSION

INTRODUCTION

Comprehension is the ability to grasp something mentally and the capacity to understand ideas and facts. Reading requires understanding, or comprehending, the meaning of the print. Readers must develop certain skills that will help them comprehend what they read and use this as an aid to reading. Comprehension skills are the ability to use context and prior knowledge to aid reading and to make sense of what one reads and hears.

Comprehension is based on :

- knowledge that reading makes sense.
- readers' prior knowledge.
- information presented in the text, and
- the use of context to assist recognition of words and meaning.

Comprehension is...

- the essence of reading.
- active and intentional thinking in which the meaning is constructed through interactions between the test and the reader (Durkin, 1973).

Definition : The complex cognitive process involving the intentional interaction between reader and the text to extract meaning is called comprehension.

Reading comprehension is the process of constructing meaning from the text. The goal of all reading instruction is ultimately targeted at helping a reader comprehend the text. Reading comprehension involves at least two people: the reader and the writer. The process of comprehending involves decoding the writer's words and then using background knowledge to construct an approximate understanding of the writer's message.

While word identification is a process that results in a fairly exact outcome (i.e., a student either reads the word "automobile" or not) the process of comprehending text is not so exact. Different readers will interpret an author's message in different ways. Comprehension is affected by the reader's knowledge of the topic, knowledge of language structures, knowledge of text structures and genres, knowledge of cognitive and metacognitive strategies, their reasoning abilities, their motivation, and their level of engagement.

Reading comprehension is also affected by the quality of the reading material. Some writers are better writers than others, and some writers produce more complex reading material than others. Text that is well organised and clear is called "considerate text". The more inconsiderate the text, the more work will be required of a reader to comprehend the text.

Reading comprehension skills separates the "passive" unskilled reader from the "active" readers. Skilled readers don't just read, they interact with the text.

Skilled readers, for instance :

1. Predict what will happen next in a story using clues presented in text.

2. Create questions about the main idea, message, or plot of the text.

3. Monitor understanding of the sequence, context, or characters.

4. Clarify parts of the text which have confused them.

5. Connect the events in the text to prior knowledge or experience.

Reading is a skill. When we read, we get exposed to the language (English) that we read. The exposure to that language enhances our communication skills, the more we read the more efficient we become. Effective and efficient reading is essential in present context. One must have the ability to read voraciously. One must also be able to remember while reading.

The important points in respect of the technique of comprehension are :

1. Good answers to the questions in the comprehension passage can be given when :

   (i) The reader understands the meaning of the passage.

   (ii) The reader properly understands the questions asked.

   (iii) He/She is able to express the meaning of the passage in his own simple and clear words.

   (iv) He must use the track of reading between the lines.

   (v) All questions have to be answered in context with the questions.

   (vi) Ability to understand the content helps in determining reliability.

   (vii) It is necessary to read a passage several times till it is completely understood.

   (viii) It is necessary to read the passage carefully atleast three times.

   (ix) It is necessary to understand the key ideas.

   (x) The reader must get the main theme of the passage i.e. what the author says about the subject matter.

2. Prepare the answers to the questions.

3. Give attention to the punctuation, grammar.

4. Prepare a gist of the passage and try to understand what the author wants to tell about the topic.

Example :

Read the following passage and answer the questions given below :

As a small boy, Edison had his laboratory in the cellar of his father's house. It contained two hundred bottles and they were all marked POISON, to keep people away. When he needed money to buy more chemicals, he managed to persuade his parents to let him seek a job. With remarkable enterprise, he obtained permission to sell newspapers on the railway train between *Port Huron* and *Detroit*.

This opportunity helped him in three further ways. First, he was able to put in a great deal of time reading at the Detroit Public Library between trains. Secondly, he thought he would start a newspaper of his own, printing it on the train and making it up from the bits of local information picked up on the line. Thirdly, without asking anyone's permission, he set up a laboratory in the van. One day, when, the train was rounding a piece of badly laid track a stick of phosphorous fell on the floor and set the van on fire. The fire was extinguished, but the angry crowd hit Edison on the ear, causing the deafness from which he suffered afterwards.

Q.1   Why did Edison mark all the bottles in his laboratory Poison ?

Ans.   Edison marked all the bottles in the laboratory 'Poison' to keep people away.

Q.2   What job did he get when he needed some money ?

Ans.   When he needed some money he got the job of selling newspapers on the train.

Q.3   Where did Edison print his newspaper ?

Ans.   Edison printed his newspaper on the train.

Q.4   Give the meaning of the following words :

Ans.   (a)   Enterprise – enthusiasm to undertake a new job.

(b)   Remarkable – extraordinary, noteworthy.

Q.5   Use the following in sentences of your own :

Ans.   1.   **To set on fire :** When the train was late, the angry crowd set the train on fire. Or, People should not set public property on fire.

2.   **Keep away :** One should always try to keep oneself away from bad habits. Or, One should try to keep oneself away from bad company.

VOCABULARY

The word vocabulary is broadly defined as a person's knowledge about words. For acquiring mastery of any language it is quite essential to increase one's word-power. In other words, more the number of words known to a person, more rich is his vocabulary. For enriching one's vocabulary there are numerous ways one can adopt. The first and foremost is to remember as many words as possible and make their frequent use in speaking and writing which helps one to remember those words with little or no efforts. Once a person gets accustomed to this practice, he or she can shift the emphasis to other aspects of increasing word power.

The further step in this direction is to get acquainted with different derivatives. By definition, derivatives are the words which are not original but obtained from other words by addition of affixes which are the meaningful elements but cannot be used independently. Affixes are either in the form of prefixes like pre -, equi -, in - or suffixes like - ment, - ish, - ly etc. For example,

pre + planned	=	preplanned
in + sufficient	=	insufficient
equi + lateral	=	equilateral
establish + - ment	=	establishment
blue + - ish	=	bluish
Friend + - ly	=	friendly

Sometimes, a word is moved from one grammatical class to another without adding a prefix or suffix. Such derivatives are called 'zero derivatives'. Derivatives are formed not only from simple words but also from derivatives. English language is so typical that there are words which carry different meanings or have different pronunciations even though their spelling is the same. Such words are called as 'homographs'. There are also words which are spelt or pronounced in the same way but carry different meanings. For example, the word 'see' is spelt and pronounced in the same way. But when used as a verb, it means 'to perceive or become aware of something by using the eyes and if used as a noun, it means' office or jurisdiction of a bishop or archbishop.

For example,

(i)   I can see the picture.

(ii)   He visited the Holy see.

Vocabulary can also be enriched by studying different words carrying the same meaning i.e. synonyms, words with opposite meanings i.e. antonyms, compound words, words that can replace a group of words, phrases, idioms, proverbs, etc. A deliberate attempt towards the above aspects of words would certainly contribute to increase one's treasure of word-power.

COMMON PHRASES

As defined in English grammar, a phrase means a combination of words that makes sense, but not complete sense. Phrases can be classified according to the parts of speech for which they are used as substitutes.

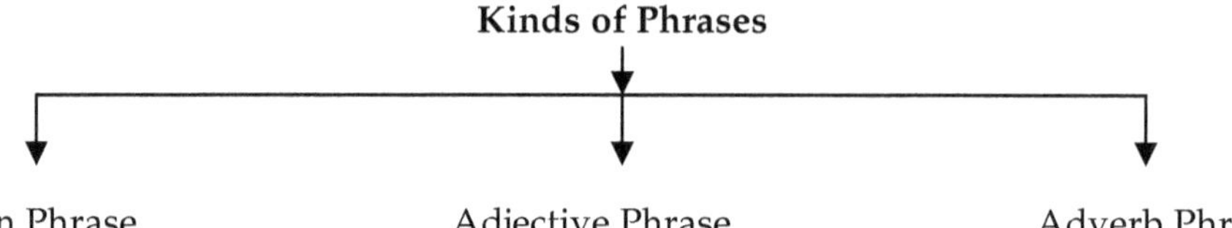

Noun Phrase : A phrase which performs the function of a noun is called a Noun Phrase.

Adjective Phrase : A phrase that does the work of an adjective is called an Adjective Phrase.

Adverb Phrase : A phrase which does the work of an adverb, is called an Adverb Phrase.

The following examples will illustrate the kinds of phrases :

(i)   **How to perform** this is a difficult task.   (Noun phrase)

(ii)   The king wears a crown **made of gold**.   (Adjective phrase)

(iii)   She cried **at the top of her voice**.   (Adverb phrase)

To acquire command on correct English, it is necessary to learn the use of phrasal verbs which are the compounds arising from very common words in English. The meanings of these verbal compounds cannot be easily understood from the mere knowledge of the normal dictionary meanings of the constituent words. It is a very common practice these days to place prepositions or adverbs after certain verbs so as to carry different meanings.

For example :

Give away   =   Give to someone/anyone.
Give up   =   Abandon.
Look after   =   Take care of.
Look for   =   Search for, seek.
Look out   =   Beware.

A student, with the knowledge of the normal meanings of the words viz., give, look, up, after, out, for etc. would be at a loss when he comes across phrasal verbs like the above formed by these words. Some of the important phrases are given below with their meaning and have been used in an appropriate sentence each :

1. **Above all** - Before every other consideration.
   Ex. Above all, you have to study hard.
2. **After all** - In spite of all that has happened.
   Ex. After all, success is the main object of all our attempts.
3. **Again and again** - Repeatedly, very often.
   Ex. It is no use stating the same thing again and again.
4. **By and by** – Before long.
   Ex. By and by he came nearer to his aim.
5. **First and Foremost** - Before anything else.
   Ex. Passing in the examination should be the first and foremost object before the students.
6. **In time** - By the proper time.
   Ex. We should go to work in time.
7. **In the long run** - In the end, eventually.
   Ex. Hard struggles do bring good results in the long run.
8. **Every now and then** - Occasionally, at odd moments.
   Ex. Every now and then they were complaining about the cost of living.
9. **On the contrary** -To the contrary, far from admiring.
   Ex. On the contrary, I have a liking for the game of cricket.
10. **Once again** - Once more, something that is repeated.
    Ex. Do not make the mistake once again.
11. **Once and for all-** Not to be repeated again.
    Ex. They settled the dispute once and for all.
12. **Over and above** - In addition to.
    Ex. Over and above the price, you have to pay 10 per cent sales tax.
13. **To and fro** - Backwards and forwards.
    Ex. He will be paid to and fro charges for his journey.
14. **At the top of** - As loud as one can.
    Ex. The students were shouting for the top of their voice in the class.
15. **Because of** - Due to, as a result of.
    Ex. He could not appear for the examination because of his ill health.
16. **By means of** - With the help of.
    Ex. The thief entered into the house by means of a rope.
17. **By the side of** - Beside.
    Ex. He was sitting by the side of the President.
18. **By virtue of** - On the strength of, On the ground of.
    Ex. He claimed seniority by virtue of his long service.
19. **For the sake of** - For the purpose of.
    Ex. He worked hard for the sake of his family.
20. **For want of** - Because of necessity.
    Ex. The building remained incomplete for want of funds.

21. **In accordance with** - With reference to, as per.
    Ex. His action is not in accordance with the rules.
22. **On behalf of** - In place of, as a representative.
    Ex. He will attend the meeting on behalf of the Principal.
23. **In case of** - For the reason of.
    Ex. I shall use these funds only in case of emergency.
24. **In consequence of** - As a result of.
    Ex. They left the village in consequence of the scarcity conditions.
25. **In the course of** - During.
    Ex. He disclosed the secret in the course of conversation.
26. **In defence of** - In support of, for strengthening.
    Ex. He resigned from service in defence of his honour.
27. **In favour of** - With inclination, with support.
    Ex. He always speaks in favour of socialism.
28. **In front of** - Before, in the presence of.
    Ex. The building was erected in front of the temple.
29. **In keeping with-** Befitting.
    Ex. His attitude was not in keeping with the occasion.
30. **In lieu of** - Substitute for, equivalent.
    Ex. I was paid a month's salary in lieu of notice.
31. **In order to** - So that.
    Ex. I had to study hard in order to pass this examination.
32. **In proportion to** - In relation to one thing to another.
    Ex. Every one has to pay income tax in proportion to one's own income.
33. **In respect of** - In point of, related to.
    Ex. He was a senior in respect of age.
34. **In spite of** - Notwithstanding.
    Ex. He made a mistake in spite of repeated warnings.
35. **Instead of** - In place of, as a substitute.
    Ex. You should keep yourself engaged in work instead of idling away your time.
36. **In view of** - Considering.
    Ex. We must make up our minds, in view of urgency.
37. **In the event of** - As a result.
    Ex. He shall have another chance in the event of failure.
38. **In the hope of** - With a desired expectation.
    Ex. He tried hard in the hope of success.
39. **On account of** - Due to, for the reason of.
    Ex. The college remained closed on account of 'Diwali' vacation.
40. **On purpose** - Intentionally, deliberately.
    Ex. It was not an accident; it was done on purpose.
41. **On the brink of** - Almost on the last point.
    Ex. The country is on the brink of disaster.
42. **On the eve of** - On certain occasion.
    Ex. The celebration was held on the eve of victory.

43. **On the ground of** - For the reason, due to.

    Ex. He declined the invitation on the grounds of his previous engagements.

44. **On the point of** - At the particular moment.

    Ex. He is on the point of breaking down.

45. **With a view to** - In order to.

    Ex. Planning was introduced with a view to achieve progress.

46. **With reference to** - Having relevance to.

    Ex. I have nothing to say with reference to your argument.

47. **At the same time** - Nevertheless.

    Ex. We accept your plan, but at the same time we want to put our own conditions.

48. **In order that** - So that.

    Ex. They made indications in order that they would be convinced.

49. **Break open-** Open something with force.

    Ex. The cupboard was broken open as the key was lost.

50. **Come true** - Turn out to be true.

    Ex. What you stated has come true.

51. **Cut short** - To bring to an end abruptly.

    Ex. His life was cut short due to the accident.

52. **Fall flat** – To fail.

    Ex. His speech fell flat on the audience.

53. **Fall short** - To become inadequate, to become less.

    Ex. It fell short of our expectation.

54. **Get rid of** - Throw it off, escape from.

    Ex. I cannot get rid of the troubles.

55. **Hold good** - Continue in force.

    Ex. This rule always holds good.

56. **Lay bare or open** - Expose.

    Ex. He managed to lay bare their secrets.

57. **Let loose** - To make free, unchain.

    Ex. He lets the animal loose at night.

58. **Look blank** – Showing no emotion.

    Ex. He looked blank when he was informed of his dismissal.

59. **Make good** - Compensate the loss.

    Ex. One cannot make good the loss of time.

60. **Make much** - To treat something as very important.

    Ex. I cannot make much of his success.

61. **Set right** - To put into working order.

    Ex. I have to set right the machine.

62. **Run short** - Become too little; less.

    Ex. This much food will run short for the invitees.

63. **Set free** - Set at liberty.

    Ex. I cannot set him free unless he accepts my conditions.

64. **Talk big** - Boast, exaggerate.

    Ex. He talks big of himself.

65. **Be against** - To oppose to
    Ex. I am against doing anything that is illegal.
66. **Be over** - Be finished.
    Ex. The storm is over; now we can go out.
67. **Blow out** - Extinguish.
    Ex. Do not blow out the candle.
68. **Blow up** - Explode.
    Ex. The enemy blew up the bridge.
69. **Break down** - Cease to function properly.
    Ex. The machine broke down when we tried to repair it.
70. **Break in** - Enter by force.
    Ex. The thieves broke in the house and took away the jewellery.
71. **Break out** - Begin, (war, fire, epidemic etc).
    Ex. An Epidemic broke out in the city.
72. **Break up** – Separate, go one's own way.
    Ex. The family broke up as all the children got married.
73. **Bring up** - Educate or train children, Raise.
    Ex. Parents work hard for bringing up their children properly.
74. **Burn down** - Destroy completely by fire.
    Ex. The oil refinery was burnt down.
75. **Call for** - Demand, require.
    Ex. The situation calls for a tactful action.
76. **Call off** - Cancel something that is not yet started.
    Ex. The strike was called off.
77. **Call out** - To summon someone to leave his house to deal with a situation outside.
    Ex. The army was called out to bring the peace.
78. **Carry on** - Continue
    Ex. I must carry on my business.
79. **Catch up** - Overtake but not pass, match, reach.
    Ex. We have to catch up to the progress made by the Westerners.
80. **Clean out** - Make the things clean and tidy, Declutter.
    Ex. I must clean out the room frequently.
81. **Clear away** - Remove articles etc. to make space.
    Ex. The table was cleared away.
82. **Clear off** - To go away from building, field, etc.
    Ex. The landlord directed the dwellers to clear off their huts.
83. **Close down** - Shut permanently.
    Ex. The factory had to be closed down on account of heavy losses.
84. **Close in** - Come nearer, approach from all sides.
    Ex. As the mist was closing in we decided to stay indoor.
85. **Close up** - Come nearer, together.
    Ex. If you close up a little, space will be available to others.
86. **Come across** - Find by chance.
    Ex. When I was cleaning my room I came across the book I had lost.

In any language, idioms play an important role. Idioms can be broadly defined as the expressions which are very peculiar to that language. Idioms in a way, decorate the language and enrich one's vocabulary.

IMPORTANT IDIOMS

1.   **A bed of roses -** an easy and comfortable situation.
     It is wrong to think that the life of an engineer is a bed of roses.
2.   **At a stretch -** without stopping to rest for a long period.
     Do you work for six hours at a stretch or get an interval in between ?
3.   **At an arm's length -** at a sufficient distance.
     We must hold bad people always at an arm's length.
4.   **At home -** feel comfortable.
     He feels quite at home with his friends.
5.   **At length -** in detail, at last.
     (i)   You must tell me the whole story at length.
     (ii) The robber was arrested at length and sent to prison.
6.   **By and by -** later on, before long.
     You will come to know every thing by and by.
7.   **In a nutshell -** in short, briefly.
     He explained the proceedings of the meeting in a nutshell.
8.   **To nip in the bud -** to check an evil in the very beginning.
     You must try to nip your evil habits in the bud.
9.   **A hair breadth escape -** a narrow escape.
     I had a hair breadth escape from being overrun by a military truck.
10. **A hard nut to crack -** difficult to solve.
     The population problem in India is a hard nut to crack.
11. **Bag and baggage -** with all personal possessions or belongings.
     He left for his native place with his bag and baggage.
12. **In black and white -** in writing.
     I do not believe in your verbal order. Please give it in black and white.
13. **By dint of -** by force of.
     He gained the first prize by dint of steady application.
14. **In full swing -** working at the highest speed, with enthusiasm.
     Preparations are in full swing for the Diwali Festival.
15. **Herculean task -** a great task.
     It is a Herculean task to conquer poverty in India.
16. **Irony of fate -** something done by evil fate/something that wasn't meant to happened but did.
     It is an irony of fate that he died.
17. **The ins and outs -** full details.
     We want such a referee who knows the ins and outs of the game.
18. **A red letter day -** an important day.
     Fifteenth August is a red letter day in the History of India.
19. **Ups and downs -** rise and fall, fluctuations in the life.
     A person who has experienced ups and downs in his life can guide others well.
20. **In the nick of time -** (just in time)
     He reached the station in the nick of time; the train was about to steam off.

EXERCISE

Use the following phrases in your own sentences to bring out their correct meanings.

1.	Come along	–	Come with me
2.	Come away	–	Leave (with me)
3.	Come in	–	Enter
4.	Come off	–	Success of a plan or scheme.
			Take place, happen as arranged.
5.	Come out	–	Exposed, Revealed, Be published.
6.	Come up	–	Rise to the surface.
7.	Crop up	–	Appear, arise unexpectedly.
8.	Cut down	–	Reduce in size or amount.
9.	Cut off	–	Disconnect, discontinue supply.
10.	Cut out	–	Cut from a piece of cloth, paper etc.
11.	Do away with	–	Abolish
12.	Drop in	–	Pay a short unannounced visit.
13.	Drop out	–	Withdraw or retire from a scheme or a plan.
14.	Fade away	–	Disappear, become gradually fainter.
15.	Fall back	–	Withdraw, retreat.
16.	Fall behind	–	Fail to keep up an agreed rate of payment or progress.
17.	Fall in	–	Get into line.
18.	Be fed up with	–	Be completely bored.
19.	Fill in/Fill up	–	To complete.
20.	Find out	–	To get some information about.
21.	Fix up	–	Arrange.
22.	Get away	–	To have a holiday, vacation.
23.	Get back	–	Recover the possession of
24.	Get over	–	Recover from illness, distress etc.
25.	Get through	–	Finish successfully.
26.	Give up	–	Abandon an attempt.
27.	Go ahead	–	Proceed, continue, lead the way.
28.	Go in for	–	Be specially interested in, practise, enter for a competition.
29.	Go through	–	Examine carefully.
30.	Grow up	–	Become an adult.
31.	Handover	–	Surrender authority or responsibility to another.
32.	Hold on	–	Wait.
33.	Keep up	–	Maintain
34.	Lay out	–	Plan gardens, building sites etc.
35.	Leave out	–	Omit.
36.	Let in	–	Allow to enter, admit.
37.	Look after	–	Take care of.
38.	Look ahead	–	Consider the future for making provision.
39.	Look into	–	Investigate.
40.	Make out	–	Discover the meaning of, understand, see, hear etc. clearly.

41.	Make up for	–	Compensate for.
42.	Pick up	–	Raise or lift a person or thing.
43.	Point out	–	Indicate or show.
44.	Pull down	–	Demolish.
45.	Put aside	–	Save for future use.
46.	Put down	–	Bring or lower down a person or a thing usually from the ground, from a table or chair/to insult a person.
47.	Ring up	–	Telephone.
48.	Run after	–	Pursue.
49.	Run away with (the idea)	–	Accept an idea too hastily.
50.	Run over	–	Drive over accidently, Overflow.
51.	See (somebody) off	–	Accompany an intending traveller to his plane/boat/ train etc.
52.	See through	–	Discover a hidden attempt to deceive.
53.	Sell out	–	Sell all that you have.
54.	Set out	–	Start a journey.
55.	Settle down	–	Become accustomed to, and contended in a new place, job etc.
56.	Show off	–	Display.
57.	Shut down	–	Close down.
58.	Stand by (someone)	–	Continue to support.
59.	Stand for	–	Represent.
60.	Take back	–	Withdraw.
61.	Take down	–	Write.
62.	Take off	–	Leave the ground (aeroplane).
63.	Take over	–	Assume responsibility.
64.	Take up	–	Begin a hobby, sport or kind of study.
65.	Think over	–	Consider.
66.	Throw away	–	Discard.
67.	Try out	–	Test.
68.	Turn down	–	Refuse, Reject an offer, application, etc.
69.	Watch out	–	Look out.
70.	Wear away	–	Gradually reduce, make smooth or flat.
71.	Wearout	–	Use till no longer serviceable.
72.	Wind up	–	Bring to come to an end.
73.	Wipe out	–	Destroy completely.
74.	Work out	–	Find by calculation or study.

CLAUSES

A clause can be defined as a *'group of words which forms a part of a sentence and contains a subject and a predicate'*. Sentences are made up of one or more clauses.

Clause Elements

There are four main elements of a clause structure. They are –

Subject	(S)
Verb	(V)
Object	(O)
Complement	(C)

For example :

He		fell		ill
Subject	+	Verb	+	Complement
(S)		(V)		(C)

Kinds of Clauses :

1. **Noun Clauses**

   We know **where we can find him** (Object of the verb know)

2. **Adverb Clauses**

   We went **where we could find him** (Adverb clause that modifies the verb **went**)

3. **Adjective Clauses**

   We went to the place **where we could find him**. (Adjective clause that qualifies the **noun place**)

Noun Clause

A noun clause is a group of words which contains a subject and a predicate of its own and it does the work of a noun.

For example : We expect **that we shall get first class.** - noun clause.

This clause in the bold letter is the object of the verb ' expect '. This is a noun clause as it does the work of a noun .

For example : That you should say this is very **strange.**

Here, the clause, 'that you should say' is the subject of the verb 'is'.

Functions of Clauses :

Two clauses in the same sentence may be related either by co-ordination or sub-ordination.

For example :

1. Ram arrived at the office by nine but no one else was there. (Co-ordination clause)

2. Ram arrived at the office by nine before any one else was there. (Sub-ordination clause).

In co-ordination, the two clauses are equal partners in the same structure, whereas in the subordination it is not the case. There is the sense of one depending on another. They are not equal in status.

Co-ordination : Ram arrived at the office by nine but **no one else was there.**

Subordination : Ram arrived at the office by ten **before anyone else was there.**

EXERCISE

Rewrite the following sentences after using noun clauses :

1. We cannot understand ....
2. We fear ....
3. I expect ....
4. Please show me ....
5. They told us ...., ...
6. Will you tell me ....?
7. Tell him ....
8. They forget ....
9. It is definite ....
10. I feel certain ....

Adverb Clause

An adverb clause is a group of words which contains a subject and a predicate of its own and does the work of an Adverb.

For example :

He may sit **where he likes -** Adverb clause.

In this sentence **'he'** is the subject and (clause **where he likes,**) is predicate and still it is a part of the sentence. Since the clause does the work of an adverb, it is called as an adverb clause.

EXERCISE

Rewrite the following sentences after using suitable adverb clauses :

1. Do not go ....
2. He was so hurried ....
3. I will miss the bus ....
4. They cannot see ....
5. The earth is larger than ....
6. He is not so intelligent ....
7. He will not go out ....
8. I sing exactly ....
9. The student went out to play ....
10. They are so busy ....

Adjective Clauses

An adjective clause is a group of words which contains a subject and a predicate of its own, and does the work of an adjective.

The car **which has white colour** is mine.

In the above sentence, the words 'which has white colour' describes the car and does the work of an adjective. It contains a subject and a predicate of its own. Hence it is called an adjective clause.

EXERCISE

Rewrite the following sentences after using suitable adjective clauses.

1. We know the place ....

2. Where is the tool ....?

3. Students .... will not be promoted.

4. He went away by the bus ....

5. The student .... won the reward.

6. We know the man ....

7. We found the book ....

8. She has lost the book ....

9. Any student .... will be punished.

10. He met a person ....

Clause Functions

In terms of function, it is necessary to study what role does a clause play in a sentence. Clauses can be first classified as subordinate and co-ordinate.

1.   Nominal Clause

Nominal clauses function mostly like noun phrases. Hence, as in the case of noun phrases, the nominal clauses also function as subject, object, complement or pre-positioned complement.

For example : That they gave a fake name shows **that** they were doing something dishonest.

In the above sentence, the two clauses introduced by the word 'that ' function as a subject and as an object respectively.

2.   Relative Clause

For example : The person who lives opposite our house is honest.

Here modifying clause introduced by the pronoun who or **that** are usually the modifiers of noun phrases. In the above sentence, the relative clause, **who live opposite our house**, modifies the noun phrase The person.

3.   Comment Clause

Such clauses comment on the truth of the sentence, the manner of saying it, or the attitude of the speaker.

Example : His plan could, I **believe**, be an important contribution towards economic progress.

Here the comment clause like "I believe" are only loosely related to the rest of the main clause. They belong to and function as sentence Adverbials. They are usually marked off from the other clause or clauses.

For example : The fillers like, 'I mean', 'I am afraid', 'you see', 'you know', 'I think' etc. are very common in informal speech.

4.   Comparative clause

Ram speaks English better than we do.

The Comparative form of adjectives and adverbs are used when it is necessary to compare one thing with another with a view to point out differences. For this purpose, a sub-clause beginning with 'than' can be introduced after the comparative word.

5.   Adverbial Clause

They have a large number of different meanings such as time, cause, contrast, reason etc.

IDENTIFYING PARTS OF SPEECH

INTRODUCTION

Parts of speech are the main constituents of traditional grammar. It is essential for the students to learn thoroughly to improve their efficiency and communication skills.

The parts of speech explain the ways words can be used in various contexts. Every word in the English language functions as at least one part of speech; many words can serve, at different times, as two or more parts of speech, depending on the context.

adjective : A word or combination of words that modifies a noun *(blue-green, central, half-baked, temporary).*

adverb : A word that modifies a verb, an adjective, or another adverb *(slowly, obstinately, much).*

article : Any of three words used to signal the presence of a noun. *'A'* and *'an'* are known as indefinite articles; *the* is the definite article.

conjunction : A word that connects other words, phrases, or sentences *(and, but, or, because).*

interjection : A word, phrase, or sound used as an exclamation and capable of standing by itself *(oh, Lord, my goodness).*

noun : A word or phrase that names a person, place, thing, quality, or act *(Fred, New York, table, beauty, execution).* A noun may be used as the subject of a verb, the object of a verb, an identifying noun, the object of a preposition, or an appositive *(an explanatory phrase coupled with a subject or object).*

preposition : A word or phrase that shows the relationship of a noun to another noun *(at, by, in, to, from, with).*

pronoun : A word that substitutes for a noun and refers to a person, place, thing, idea, or act that was mentioned previously or that can be inferred from the context of the sentence *(he, she, it, that).*

verb : A word or phrase that expresses action, existence, or occurrence *(throw, be, happen)*. Verbs can be transitive, requiring an object *(her in I met her)*, or intransitive, requiring only a subject (The sun rises). Some verbs, like feel, are both transitive *(Feel the fabric)* and intransitive *(I feel cold, in which cold is an adjective and not an object)*.

> **All the above given parts of speech are discussed in detail with examples and exercises in Chapter 2.**

Part – II

APPLICATION OF GRAMMAR

(1) Verbs
(2) Tenses
(3) Do as Directed
    (A) Active/Passive
    (B) Direct/Indirect
    (C) Affirmative/Negative
    (D) Assertive/Interrogative
    (E) Question Tag
    (F) Remove Too
    (G) Use of Articles
    (H) Prepositions
    (I) Conjunctions
    (J) Interjections
    (K) Punctuation
(4) Correct the Errors from the Sentences

[Marks 18]

Language in its widest sense means the sum total of such signs of our thoughts and feelings as are capable of external perception and could be produced and repeated at will. It can also be stated as 'the expression of thoughts' by means of speech sounds. It is an organisation of sounds of vocal symbols - the sound produced from the mouth to convey some meaningful message. It also means that speech is primary to writing.

In any language, the smallest unit of full expression is a sentence. It consists of one or more utterances arranged according to the specific system of that particular language. The specific system of any language concerned with the written and spoken form of words and sentences is broadly termed as 'grammar' of that language.

For any student studying language, it is necessary that he should be conversant with the basic grammar of that language. When we think of English language, students must be aware about the basic concepts of grammar and the usages of different parts of speeches. The study of any language begins with the letters of alphabet and combination of these alphabets to form words, phrases, clauses and sentences. Sentences are the basic elements from which we get passages of prose, poetry and other forms.

A written form is essentially a representation of the spoken form. However, we may not always find exact reflection of spoken form in the written representation. For example, the letters 'ought' have different spoken forms in the words like *though, rough, bought, bough, dough*, etc. Thus, it is necessary to understand that there is a difference in the writing form and the spoken form. Besides, even the written form sometimes differs in the same language. In British system, the word is spelled as 'colour', while in the American system of writing English, it is 'color'. Hence, it can be said that writing often fails to represent the sounds fully and accurately.

(2.1)

The traditional grammar works with two main units of grammatical description viz. the word and the sentence. Both these units are given a practical recognition in the writing system. In this system, sentence is the highest unit while morpheme is the lowest one. The different units of writing can be arranged in the scale of rank as (i) sentence, (ii) clause, (iii) phrase, (iv) word, (v) morpheme. The higher rank is composed of the lower rank. Certain important features of these units can briefly be stated as below :

Sentence

A sentence is by definition and conveniently taken as the largest unit of the grammatical analysis. It is the upper limit of structural statement at the grammatical level.

Features of a sentence are as follows :
1. Sentence is preceded and followed by infinite pause or silence;
2. It has the phonetic features associated in each language with pre-pausal position;
3. It is usually marked in writing by final punctuation mark such as full stop (.), question mark (?), exclamation mark (!) or semi colon (;).
4. It has an intonation tune in speech;
5. There is always a pause at the sentence boundary;
6. In speech, it may be uttered with a pre-pausal intonation tune;
7. Sentence expresses either a complete predication, question or command and each has a specific logical form;
8. The predicative sentence is often equated with the position of logic;
9. Grammar operates between the upper limit of the sentence and the lower limit of the morpheme;
10. Between the ranks viz. sentence and a morpheme we have such ranks as clause, phrase and word.

Clause

A group of words with its own subject and predicate, included in a large sentence is called as a clause. The phrase and the clause are secondary grammatical units.

Phrase

A phrase can be defined as 'any group of words which is grammatically equivalent to a single word and does not have its own subject and predicate.

The Word

The term 'word' can be defined as 'speech utterance', or 'verbal expression'. Word is the smallest unit of any written discourse. The word is a very important fundamental unit.

Certain features of word are as follows :
1. A word possesses a single meaning;
2. It conveys a single idea;
3. It is minimum free form of a language;
4. The word as a stretch of speech admits minimal pause on either side;
5. Words are constituted out of morphemes;
6. How a word can be divided into the smaller grammatical segments is a matter of degree. For example, the words like 'girl-'s 'call'-s, call-ing, tall-er, tall-est etc. can be segmented into their constituent parts which may not be readily acceptable.
7. Words are determinate with respect to segmentation;
8. There are words which cannot be segmented. For example: men, teeth, nice, better, best, worse etc.
9. Morphology deals with the internal structure of words and syntax with the rules governing their combination in a sentence.
10. The words are represented differently. For example, the phonological form /kæt/ has the orthographical form <cat>

The corresponding orthographic word 'cut' represents three grammatical forms :

(a)   Present tense - cut   (b) Past participle – cut   (c)   Past tense - cut

11.   The word can be taken as the union of a particular meaning with a particular complex of sounds of a particular grammatical employment.

Morpheme

The morpheme is the minimum grammatical unit. For example there are four constituents 'un-faith-ful-ness' in the word unfaithfulness. These constituents are called morphemes.

In each language, there are certain permissible ways of organising morphemes. The constituents of a sentence are nothing but the morphemes or a group of morphemes which, when structured into successive components, form utterances (sentence).

For example : " The poor man ran away ".

If we wish to split up this sentence into two immediate constituents, that will become –

The poor man and ran away . If we try to split it into still further constituents, it will become

The poor and man , and ran and away . So, a sentence is seen not as a sequence or 'string ' of elements like The + Poor + man + ran + away. It is made up of 'layers' of constituents, each cutting 'points' or 'note' .

Example :

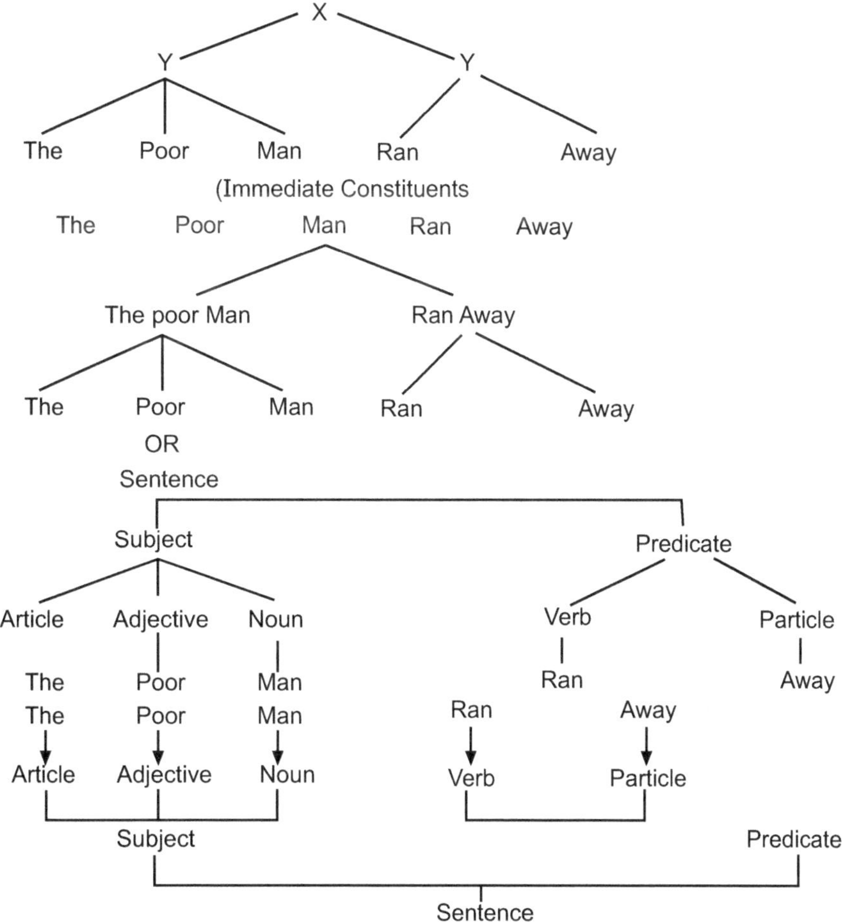

The above sentence is made up of five morphemes, which may be defined as the minimum significant syntactic units :

The   Poor   Man   Ran   Away

1   2   3   4   5

These elements are not further analysable at the syntactic level. These constituents are organised in a particular order in the sentence. A jumble of morphemes thrown together at random might produce non-sentence like :

* Away Man Poor Ran The

Hence, each human language has certain permissible ways of organising morphemes to make them a sentence. The sequential ordering of the ultimate constituents shown above is called the sentence. Thus, morphemes, words, phrases and clauses are all constituents of a sentence though all of them are not ultimate constituents.

Words are divided into different classes called 'Parts of Speech' according to their use. In English language we generally speak of eight parts of speech. They are noun, pronoun, adjective, adverb, verb, preposition, conjunction and interjection. In the traditional grammar, the major parts of speech were associated with certain typical functions in a simple sentence.

NOUN

A noun is a word used as the name of a person, place, or thing. The word **thing** includes all that which we can see, hear, taste, touch or smell i.e. something which we can think of. For example :

> The **sun** shines brightly.
>
> His **courage** rewarded him.
>
> The **flower** smells sweet.

In the above sentences, the words in bold type are nouns.

Nouns can be further classified according to their kinds. They are as follows :

NOUN

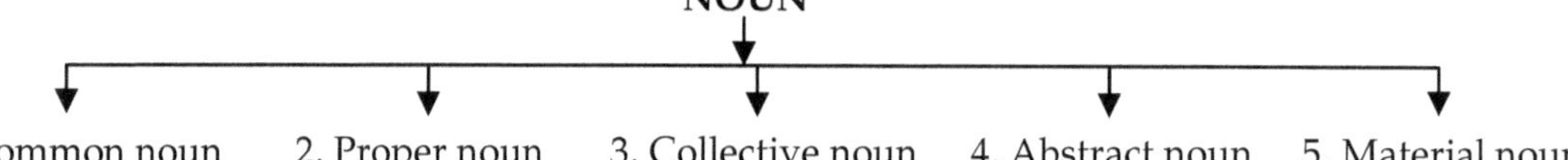

1. Common noun   2. Proper noun   3. Collective noun   4. Abstract noun   5. Material noun

Common Noun

A Common noun is a name given in common to every person or thing or a unit of the same class or kind. Common means it is shared by all. For example, A **boy** is playing a game.

In this sentence, the word 'boy' is a common noun because it applies to any boy.

Proper Noun

A Proper noun is a name of some particular person or place. Here it means one's own name. Proper nouns are written with their first letters in capital. For example,

> **Sita** is playing a game.

In the above sentence, the word Sita is a proper noun and the first alphabet of the words are written in a capital letter.

Collective Noun

A Collective noun is the name of a number (or collection) of persons or things taken together and spoken of as one whole. For example,

> The **army** is advancing ahead.
>
> They dispersed the **crowd.**

In the above sentences, the words in bold letters i.e. army and crowd, represent collection of persons as one whole. Similarly, the words like fleet, class, jury, police, battalion are taken together and represent the collection as one whole.

Abstract Noun

An Abstract noun is usually the name of quality, action or state considered apart from the object to which it belongs. Take for example the sentences like,

We have to test the **hardness** of the metal. They have showed **bravery**.

In these sentences, the words 'hardness' and 'bravery' are abstract nouns which denote the mental quality or the characteristics of persons.

Other examples of abstract noun are, goodness, kindness, whiteness, darkness, brightness, honesty, wisdom, laughter, theft, movement, judgement, hatred, childhood. boyhood, youth, slavery, sleep, sickness, death, poverty, art, music, grammar, chemistry etc. All these words come under the category of Abstract noun.

Material Noun :

A Material noun is the name of material. For example,

Sugar production is considerably increased.

Quality **wood** is required for the furniture.

In the above sentences; the words 'sugar' and 'wood' are material nouns.

EXERCISE

Use the following nouns in a sentence of your own classifying them into Proper, Collective, Abstract or Material nouns.

Cattle, soldiers, sailors, cruel, foolish, oil, lazy, rice, truth, class, strength, cleanliness, fleet, the Ganges, committee, happiness, team, wisdom, room, childhood.

Gender

A noun that shows a female animal is said to be **Feminine Gender** and the one that denotes male is said to be **Masculine Gender**.

A noun that shows either male or female is said to be of **Common Gender**.

A noun that shows a thing that is neither male nor female is said to be **Neuter Gender**.

These are three ways to change Masculine Nouns into Feminine Nouns :

1. By using altogether different word

Examples :	Masculine	Feminine
	Boy	Girl
	Bull	Cow
	Cock	Hen
	Father	Mother
	Man	Woman

2. By adding syllables like — 'ess' —'ine' —'trix' — 'a' etc. at the end

Examples :	Masculine	Feminine
	Author	Authoress
	Lion	Loiness
	Poet	Poetess
	Actor	Actress

(In certain cases last vowel of the word is dropped)

	Tiger	Tigress
	Master	Mistress
	Hero	Heroine
	Sultan	Sultana

3.   By placing a word before or after

Examples :	**Masculine**	**Feminine**
	He goat	She goat
	Man servant	Maid servant
	Milkman	Milkmaid
	Peacock	Peahen

The Noun - Number

A noun which shows one person or thing is said to be in the singular number and the noun which shows more than one person or thing is said to be in the plural form.

Formation of Plurals

1.   Plurals of nouns are generally formed by adding suffix 's' to the singular noun.

Examples :	**Singular**	**Plural**
	Boy	Boys
	Pen	Pens
	Book	Books
	Room	Rooms

2.   In some cases, plurals are formed by adding 'es' at the end of the word where the noun ends in '-s'—'sh'—'ch' or 'x'

Examples :	**Singular**	**Plural**
	Class	Classes
	Branch	Branches
	Box	Boxes
	Bush	Bushes
	Tax	Taxes

3.   In most of the cases of nouns ending in '– o', plurals are obtained by adding 'es' to the singular

Examples :	**Singular**	**Plural**
	Potato	Potatoes
	Mango	Mangoes
	Cargo	Cargoes
	Echo	Echoes

4.   In some cases, however, a noun ending with '– o' takes only 'S'

Examples :	**Singular**	**Plural**
	Dynamo	Dynamos
	Photo	Photos
	Ratio	Ratios

5.   Nouns that end in 'y' preceded by a consonant, form their plurals by changing '– y' into '– i' and adding 'es' in the end.

Examples :	**Singular**	**Plural**
	Baby	Babies
	Army	Armies
	Lady	Ladies
	City	Cities

6.   Nouns that end in — 'f' or — fe' form their plurals by changing — 'f, or —'fe' into 'v' and 'es' in the end.

Examples :	Singular	Plural
	Thief	Thieves
	Life	Lives
	Knife	Knives
	Leaf	Leaves
	Self	Selves

Certain cases are exception to this note.

Examples :	Singular	Plural
	Chief	Chiefs
	Roof	Roofs
	Grief	Griefs
	Proof	Proofs

7.   Certain nouns form their Plural by changing the inside vowel of the singular noun.

Examples :	Singular	Plural
	Man	Men
	Woman	Women
	Foot	Feet
	Tooth	Teeth

8.   In certain cases, nouns have the same form as their plural form.

Examples :	Singular	Plural
	Sheep	Sheep
	Hundred	Hundred
	Dozen	Dozen
	Fish	Fish

9.   Some nouns are used only in the plural form.

Examples : Spectacles, tongs, scissors, trousers, thanks, assets etc.

10.   Certain plural nouns are commonly used as singular.

Examples : Mathematics, news, mechanics, politics, innings.

11.   Some nouns originally singular, are generally used as plural.

Examples : Riches, Alms.

12.   Certain collective nouns, singular in form are always considered as plurals.

Examples : Poultry, cattle, people etc.

(When people means nation, it is both singular and plural).

They are hardworking and brave people.

There are many different people in the country.

13.   Generally, plural of a compound noun is formed by adding 's' to the root word.

Examples :	Singular	Plural
	Commander-in-chief	Commanders-in chief
	Son-in-law	Sons-in-law
	Step-son	Step-sons
	Maid-servant	Maid-servants
	Passer-by	Passers-by

14. Nouns which are taken in English language from other foreign languages keep their original plural form.

Examples :	**Singular**	**Plural**
	Index	Indices
	Radius	Radii
	Axis	Axis
	Basis	Basis
	Analysis	Analysis
	Hypothesis	Hypothesis
	Phenomenon	Phonomena
	Criterion	Criteria
	Madame	Madames

15. Some nouns have two forms for their plural, with different meaning.

Examples :	**Singular**	**Plural**
	Cloth	Clothes (garments)
		Cloths (pieces of cloth)
	Index	Indexes (tables of contents of books)
		Indices (signs used in Algebra)

16. Certain nouns have two meanings in the singular form but only one in the plural.

Examples :	**Singular**	**Plural**
	1. Radiance	
Light		Lights (lamps)
	2. A lamp	
	1. Habit	
Practice		Practices (habits)
	2. Exercise of a profession	

17. Certain nouns have one meaning in the singular form and two in the plural form.

Example :	**Singular**		**Plural**
			1. habits
	Custom (habit)	Customs :	
			2. duties levied on imports
			1. results
	Effect (result)	Effects :	
			2. property
			1. fourth parts
	Quarter (fourth part)	Quarters :	
			2. lodgings
			1. closed land attached to home
	Ground 1. Dearth	Grounds :	
	2. Reasons		2. reasons.

18. Certain nouns have different meanings in the singular and in the plural.

Examples :	**Singular**	**Plural**
	Advice (counsel)	Advice (information)
	Good (benefit, well being)	Goods (merchandise)
	Respect (regard)	Respects (compliments)
	Force (strength)	Forces (troops)

Nouns

1.   **Use the following nouns in sentences of your own.**

<table>
<tr><td colspan="4" align="center">Exercise</td></tr>
<tr><td>Amateur</td><td>Graft</td><td>Mechanism</td><td>Self-sufficiency</td></tr>
<tr><td>ambition</td><td>habitate</td><td>Mentor</td><td>Smog</td></tr>
<tr><td>Ancestors</td><td>husk</td><td>Metaphor</td><td>Stability</td></tr>
<tr><td>Astronaut</td><td>Illustrations</td><td>Nickel</td><td>Stage</td></tr>
<tr><td>Championships</td><td>Imagination</td><td>Novice</td><td>Suspicion</td></tr>
<tr><td>Democratization</td><td>Impropriety</td><td>Peasants</td><td>Suspicion</td></tr>
<tr><td>Distaste</td><td>Impudence</td><td>Persuasion</td><td>Synthesis</td></tr>
<tr><td>Electrification</td><td>Initiative</td><td>Radiator</td><td>traverses</td></tr>
<tr><td>Environment</td><td>Loveliness</td><td>Rescue</td><td>Violence</td></tr>
<tr><td>forest</td><td>Maturity</td><td>Retrieval</td><td>Vocation</td></tr>
<tr><td>Fraction</td><td>Mechanisation</td><td>Salvation</td><td>Windbag</td></tr>
</table>

Pronoun

A word which is used instead of a noun is called a pronoun. A pronoun can function as a whole noun phrase, as a subject or an object of a clause. Many of them (pronouns) act as a substitute or replacement for a noun phrase in the context. Many words can function both as determiners and as pronouns.

For example :   Which instrument is yours ? (Here the word 'which' is a determiner)

Which is yours ? (Here the word 'which' is a pronoun)

This book is mine. (Here the word 'this' is a determiner)

This is my book. (Here the word 'this' is a pronoun)

Forms of Pronouns

1.   Demonstrative pronouns (this, that, those, these)
2.   Interrogative pronouns (who, which, what, where etc.)
3.   Negative pronouns (none, nobody, no one, nothing etc.)
4.   Personal and reflexive pronouns (I, my, mine, myself)
5.   Reciprocal pronouns (each other, one another etc.)
6.   Quantifiers (some, any, someone, everything, anybody, each, all, both, either, much, many, more, most, enough, several, little, few, less, least etc.)
7.   Relative pronoun (who, whom, whose, which, that etc.)

Personal and Reflexive Pronouns

Personal pronouns and reflexive pronouns are related.

Below the table gives all the forms of personal and reflexive pronouns.

		Personal pronouns		Possessive		Reflexive pronoun
		Subjective	Objective	Acting as determiner	Acting as pronouns	
1st Person	Singular	I	me	my	mine	myself
	Plural	we	us	our	ours	ourselves
2nd Person	Singular and Plural	you		your	yours	yourself
3rd Person	Singular Masculine	he	him	his		himself
	Singular feminine	she	her	her	hers	herself
	Singular non-personal	it		its		itself
	Plural	they	them	their	theirs	themselves

Personal pronouns are classified according to person, number, gender and case.

Important uses of a Pronoun

1.   A Pronoun can be used for things without life.

For example : Here is your drawing board take **it** away.

2. A Pronoun can be used for animals, unless we clearly wish to speak of them as **male or female**.

   For example : They love their animal and cannot do without **it**.

3. A Pronoun can be used for a young child.

   For example : When we observed the child **it** was sleeping.

4. A Pronoun can be used to refer to some statement going before.

   For example : They deserved their reward, as they knew **it**.

5. A Pronoun can be used as a provisional and temporary subject before the verb **to be** when the real subject follows.

   For example : It is easy to find a mistake (to find a mistake is easy).

   It is doubtful whether they will come.

6. A Pronoun can be used to give emphasis to the noun or pronoun following.

   For example : It was she who began the quarrel.

   It was I who first objected.

7. A Pronoun can **be** used as an indefinite nominative of an impersonal verb.

   For example :   **It** rains, **It** thunders.

   Here the pronoun **'it'** does not represent any noun whatsoever, though this can be supplied from the verb. Here it means **rains**.

8. A Pronoun can be used in speaking of weather or time.

   For example :   It is five o'clock. It is hot. It is fine.

Since a personal pronoun is used instead of a noun, it must be in the same number, gender and person as the noun it represents.

EXERCISE

Rewrite the following sentences using correct form of pronouns.

1. It isn't fair for .... to dictate to me. (they, them)
2. Nobody but .... was present. (He, him)
3. She is known to my father and .... (I, me)
4. We are not so rich as .... (they, them)
5. She is as old as .... (I, me)
6. Let .... answer this. (He, him)
7. He and .... were known friends. (I, me)
8. He is as strong as .... (I, me)
9. We scored as many marks as .... (they, them)
10. Hari and .... were absent (I, me)

---

1. VERBS

Verb is the most important word in a sentence. It is a word which tells something about a person or thing i.e. the subject in the sentence. It indicates :

1. What a person or thing does;
2. What is done to a person or thing;
3. What a person or thing is.

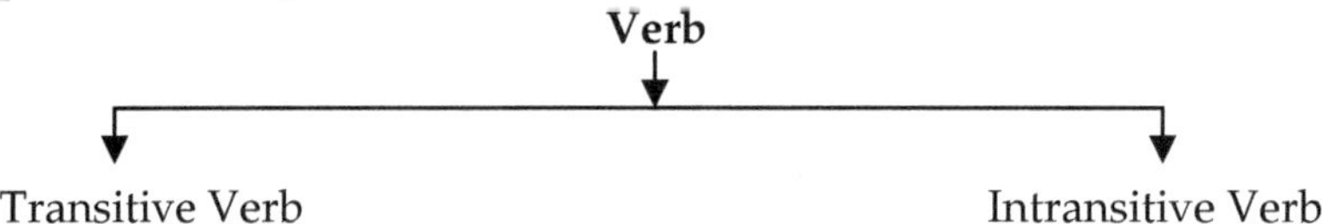

Transitive verb

When the action denoted by the verb **passes** from subject to some object, the said verb is called a transitive verb.

For Example : 1. He **played** cricket. 2. He **wrote** a letter.

In the above sentences, the action denoted by the verbs passes over from the doer (subject) to some object. Therefore, these verbs are called as transitive verbs.

Intransitive Verb

When the action denoted by a verb stops with the doer or subject and does not pass over to an object, the said verb is called as an ' Intransitive Verb. '

For example : 1. She laughs loudly. 2. He runs.

1. Most transitive verbs take a single object. But there are some verbs like ask, offer, give etc. which take two objects after them.

   For example : (i) The teacher gave him (Indirect) a reward (direct).

   (ii) He disclosed to me (Indirect) a secret (direct).

2. Most of the verbs are used both as Transitive and Intransitive. It can therefore, be, said that a verb is used Transitively and Intransitively.

   For example : (i) The driver **stopped** the car. (used 'Transitively')

   (ii) The car **stopped** at once. (used 'Intransitively')

3. Certain verbs like, come, go, sleep etc. denote actions which cannot be done to anything. Such verbs are Intransitive.

4. In case the subject and the object both refer to the same person, the verb is said to be used reflexively.

   For example : The person **killed** himself.

   Please **keep** quiet.

5. Certain verbs can be used reflexively and also as ordinary transitive verbs.

   For example : (i) Do not forget my name (Reflexive)

   (ii) I forgot his name. (Transitive)

   (iii) I enjoy myself alone. (Reflexive)

   (iv) She enjoys good health. (Transitive)

EXERCISE

Use the following verbs as transitive and intransitive.

1. Walk 2. Run 3. Fly

---

2. TENSES

Verb occupies the most significant place among all parts of speech, as it is the backbone of any sentence which should make sense and serve the basic purpose of correct expression of thought or action. It is, therefore, quite essential for the students to understand thoroughly the method of using a verb in a sentence.

In English grammar, verb denotes an action or event. It also indicates the time of the action and the state of the action. On the basis of these two aspects, the verb takes its exact form to indicate the time and the degree of completeness attributed to the action. Thus, according to these factors, the verb may be in

1. Present time 2. Past time 3. Future time.

According to the time factor, there are three main tenses.

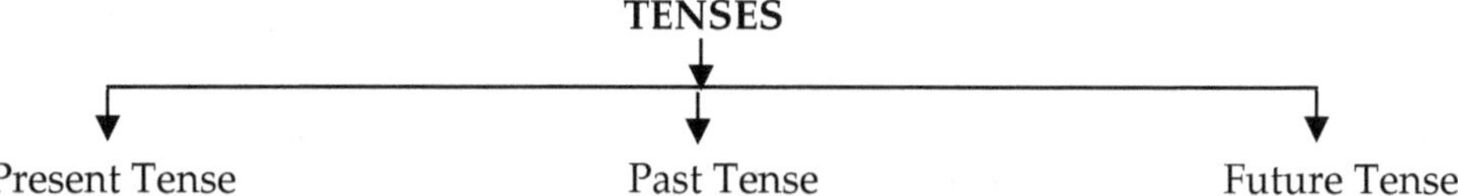

1. A verb used for denoting an action in the present time is said to be in the Present Tense.

2. A verb used for denoting an event or an action taken place in the past time is said to be in the Past Tense.

3. A verb used for denoting an event or an action that will take place in the time to come is said to be in the Future Tense.

For example :
    (i)   He writes a letter.   –   Present Tense
    (ii)   He wrote a letter.   –   Past Tense
    (iii)   He will write a letter.   –   Future Tense

In the above sentences, although the action is the same i.e. of writing a letter, it refers to different times. In the first sentence it refers to the present time. The verb of this sentence is, therefore, in the present tense. The second sentence states that the action of writing a letter took place sometime in the past. As such, the verb in this sentence is in the past tense. However, the verb in the third sentence refers neither to the present nor to the past but to the action that is yet to take place or will take place sometime after the present time i.e. in future. As such, the verb in the third sentence is in the future tense.

Each of these three tenses has four forms based on the state of action or condition of action. The state or condition of action is that aspect of a verb which explains the degree of completeness of an action i.e. whether the action is :

1. Indefinite (simple without indicating completeness or incompleteness of the action).

2. Imperfect (incomplete or continuous going on).

3. Perfect (completed, finished or perfect at the time of speaking).

4. Perfect continuous. (Part of the action completed at the time of speaking and would continue in future also).

Thus, according to the state of action, each tense is further divided into four forms viz.

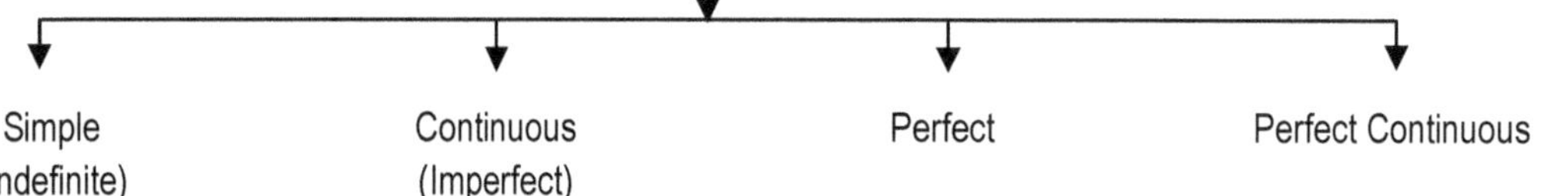

1. When an action is simply mentioned without anything being said about the completeness or incompleteness of it, the form of the verb is called Indefinite.

    For Example :
    1.   He writes a letter   – – –   Present Indefinite.
    2.   He wrote a letter   – – –   Past Indefinite
    3.   He will write a letter   – – –   Future Indefinite.

2. When an action is spoken of as going on at a point of time either in the present or in the past or in the future, the form of the verb is said to be in the continuous or Imperfect. This means that the verb shows that the action mentioned is incomplete, and is still going on at the time of speaking and hence imperfect.

Consider the following examples :

1.	He is playing a game	– – –	Present Continuous or Imperfect.
2.	He was playing a game	– – –	Past Continuous or Imperfect.
3.	He will be playing a game	– – –	Future Continuous or Imperfect

When an action is completed just at the time of speaking or perfect either in the present or in the past or in the future, the form of the verb is called perfect.

4. When an action is a combination of present, past and future it is said to be a Perfect continuous one. This can be better explained through an example :

I have been writing a letter

In this sentence, the action of writing letter has already started at a point of time before the moment of speaking; part of it is completed at the time of speaking and shall be further continued in the time to come i.e. in future. This is the present perfect continuous. It is applicable in the same sequence for all tenses viz. past and future perfect continuous tenses.

For Example :

(a)	I have written a letter	_ _ _	Present Perfect.
(b)	I had written a letter	_ _ _	Past Perfect.
(c)	I shall have written a letter	_ _ _	Future Perfect.

In the above sentences, the verb shows that the action of writing a letter is finished just at the time of speaking (Completed) either in the present or in the past or in the future.

For example,

I had been writing a letter	_ _ _	Past Perfect Continuous.
I shall have been writing a letter	_ _ _	Future Perfect Continuous.

For example :

1.	I have been writing a book	_ _ _	Present Perfect Continuous.
2.	I had been writing a book	_ _ _	Past Perfect Continuous.
3.	I shall have been writing a book	_ _ _	Future Perfect Continuous.

1. The Present Simple or Indefinite Tense :

In the present simple or Indefinite tense, the verb shows that the action mentioned in the sentence indicates habit. Without being said anything about its completeness or otherwise.

I	play	a	game
We	play	a	game
You	play	a	game
He	plays	a	game
She	plays	a	game
It	plays	a	game
They	play	a	game

Here you can notice that in case of the sentence using third person singular number as a subject, 's' or 'es' is added to the main verb. From the above sentences, it can be noted that the sentence in the present simple tense has a particular structure as given below :

Subject + Main Verb (Transitive) + Object

If we remember the structure of the sentence in this tense, we can form any number of sentences of this type.

Usages : The Simple Present or which is also known as Present Indefinite tense is used :

(a) To express what is actually taking place at the present moment.

For example :   Rama plays with his toys.

See, how it falls.

(b)   To express universal truths :
      For example :   The earth is round.
                            Honesty is the best policy.
(c)   To express a habitual action.
      For example :   Hari drinks tea daily.
                            We exercise in the evening.
(d)   To express futurity when it is indicated by context ; (instead of using future tense).
      For example :   I plan to go to Mumbai next week.
(e)   To express vivid narrative as a substitute for the past tense. (This type is also called the
      Historic present).
      For example :   Shivaji escapes from Agra and comes back to his capital in disguise.

2.   The Present Continuous or Imperfect Tense

In the Present Continuous Tense, which is also called as the present imperfect tense, the verb shows continuity of the action mentioned. That is to say that the action is going on or incomplete or is in progress at the time of speaking.

I	am	writing	a	book
We	are	writing	a	book
You	are	writing	a	book
He	is	writing	a	book
She	is	writing	a	book
It	is	writing	a	book
They	are	writing	a	book

From the above examples, it can be seen that the structure of sentences in the Present Continuous Tense is as given below :

Subject + is/am/are + Main Verb (in the present participle form i.e. main verb + ing) + object.

Usages : The present continuous tense is used to express :
(a)   An action going on at the time of speaking.
      For example : He is going home.
(b)   An action which is to take place in the near future.
      For example : I am leaving for Mumbai tomorrow.

3.   Present Perfect Tense :

The present perfect tense shows the action started sometime before in the past and just finished at the time of speaking :

I	have	written	a	letter
We	have	written	a	letter
You	have	written	a	letter
He	has	written	a	letter
She	has	written	a	letter
It	has	written	a	letter
They	have	written	a	letter

In the present perfect tense, auxiliary verb 'to have' is used as a verb helping to the main verb. From the examples given above, it will be seen that the structure of sentences in the present perfect tense is as below :

Subject + have/has + Main verb (in the past participle form) + Object

Usages :

The present perfect tense is used to express an action started in the past and has just been completed. It is also used instead of the past tense to represent past action which it refers to present.

For Example : We have lived here for ten years.

　　　We lived here for ten years.

In the first sentence, the sentence denotes that we are likely to live here in the future, whereas the second sentence, denotes that we are no longer living here.

4.　Present Perfect Continuous Tense :

In the present perfect continuous tense, the verb shows that the action referred is started in the past, partly completed at the time of speaking and still continues in the future. The present perfect tense carries the sense of a single action completed at the time of speaking; whereas, in the use of the present perfect continuous tense, there is a sense of continuity or that the action is prolonged in the future.

I	have	been	writing	a	letter
We	have	been	writing	a	letter
You	have	been	writing	a	letter
He	has	been	writing	a	letter
She	has	been	writing	a	letter
It	has	been	writing	a	letter
They	have	been	writing	a	letter

From the above examples, it will be seen that the structure of sentences in the present perfect continuous tense can be represented in the following ways :

Subject + have/has + been + Main Verb (in the Present participle form i.e. verb + - ing) + Object. (ing)

Past Tense :

(1)　The Past Simple or Past Indefinite Tense : In the past simple tense the verb shows that the event was completed in the past and thus it is separated from the present.

I	wrote	a	letter
We	wrote	a	letter
You	wrote	a	letter
He	wrote	a	letter
She	wrote	a	letter
It	wrote	a	letter
They	wrote	a	letter

From the above sentences, it can be seen that the structure of a sentence in the past simple or indefinite is given as follows.

Subject + Main verb in the past tense form + object.

Usages :

The past simple or indefinite tense is used to express :

(a)　An action that took place and completed in the past.

　　　(i)　She went home. (ii)　They saw a picture.

(b)　Sometimes an action in the past continuous tense can also be expressed in simple past tense.

　　　He played (i.e. was playing) while I watched (i.e. was watching).

In this case it means that the use of simple past tense is made instead of past continuous tense.

(c)   To express habitual action in the past.

He took tea daily.

(2)   The Past Continuous Tense : The past continuous tense represents the action which was going on at a point of time in the past.

I	was	writing	a	letter
We	were	writing	a	letter
You	were	writing	a	letter
He	was	writing	a	letter
She	was	writing	a	letter
It	was	writing	a	letter
They	were	writing	a	letter

From the above sentences, it can be seen that the structure of a sentence in the past continuous tense is as given below :

Subject + was/were + Present participle form of the Main Verb

(i.e. verb + ing) + Object

Usages : The past continuous tense is used to denote an action that was in progress or going on at a point of time in the past.

(3)   Past Perfect Tense : The past perfect tense which is also known as the plu-perfect tense denotes an action completed at a point of time in the past to which the context relates. The past perfect tense is used only when we have to show priority of one past event over another past event. Where no such priority of one event over another event is implied, the past indefinite tense should be used.

The structure of sentences in the Past Perfect Tense is as follows :

Subject + had + Main Verb (in the past participle form) + Object

Subject	Had	Main Verb in past participle form	Object	+ Clause for indicating priority
I	had	drunk	the milk	before I went to bed
We	had	drunk	the milk	before we went to bed
You	had	made	a noise	even before the dog entered the room.
Rama	had	played	cricket	before he joined school.
She	had	sung	a song	before the guest arrived
They	had	read	a book	before it was banned.

Usages : The past perfect tense is used to denote an action completed at some point of time in the past before another action commenced.

For example : I had written a letter, before he came back.

In the above example, two actions in the past i.e.

(i)   Writing a letter and

(ii)   Coming of a person, are relatively mentioned.

One action has a context with the other. The action which preceds (writing letter) another action (coming), is expressed in the Past Perfect Tense while the latter action is expressed in the Simple Past Tense.

(4) Past Perfect Continuous : The past perfect continuous tense represents an action that –

(i)   Started some time in the past.

(ii) Partly completed at a point of time in the context and had been continued beyond the time mentioned.

I	had	been	writing	a	letter
We	had	been	writing	a	letter
You	had	been	writing	a	letter
He	had	been	writing	a	letter
She	had	been	writing	a	letter
It	had	been	writing	a	letter
They	had	been	writing	a	letter.

The above examples indicate that the action referred in the sentence has a reference to a particular time in the past when the action of writing of a letter already started and was partly completed before the particular moment. The structure of sentences in the past perfect continuous tense is as given below.

Subject + had + been + Main Verb (ing) + Object.

Usages : The past perfect continuous tense is used whenever we want to state an action that started in the past, partly completed at a point of time referred in and further continued for further period in the past.

Future Tense : Future tense of all verbs is formed with the help of the auxiliary verbs 'shall' and 'will'. 'Shall' is used when the subject is in the first person while 'will' is used when the subject is in the second and third person. Consider the use of this tense with the personal pronouns :

I/We	shall	speak
You	will	speak
He/She/It	will	speak
They	will	speak

Thus, the use of "shall" with the first person and "will" with the second person and third person indicates simple futurity of the action. However, the use of "will" with the first person and 'shall' with the second and third person expresses emphasis or determination of the speaker to perform the action.

Person	For expressing mere futurity	Emphasis determination obligation
First	I shall speak.	I will speak.
Second	We shall speak	We will speak
	You will speak.	You shall speak.
	You will speak	You shall speak
Third	He/she/it will speak.	He/she/it shall speak.
	They will speak.	They shall speak.

1. **Future Indefinite Tense :** The future Indefinite tense represents an action that will take place in future.

I	shall	write	a	letter
We	shall	write	a	letter
You	will	write	a	letter
He	will	write	a	letter
She	will	write	a	letter
It	will	write	a	letter
They	will	write	a	letter

The structure of a sentence in the future Indefinite tense can be represented as :

Subject + shall/will + Main Verb + object.

Usages : Future Indefinite Tense is used to denote an action which is to take place at a point of time in future.

2.	**The Future Continuous Tense :** The future continuous tense represents an action which shall be going on or would be in progress at a certain point of time in the future.

I	shall be	writing	a	letter
We	shall be	writing	a	letter
You	will be	writing	a	letter
He	will be	writing	a	letter
She	will be	writing	a	letter
It	will be	writing	a	letter
They	will be	writing	a	letter.

The structure of sentences in the future continuous tense is as follows :

Subject + shall/will + be + Main Verb (ing) + Object.

Usages : The future continuous tense is used to describe an action that will be going on at some point of time in the future that is to be finished at a particular time in future.

It is also used to express two future actions one of which will be completed earlier than the other.

For Example : He will have reached there before you start.

3.	**The Future Perfect Tense :** The future perfect tense represents an action that will be completed at a particular point of time in the future.

I	shall	have	played	a	game
We	shall	have	played	a	game
You	will	have	played	a	game
He	will	have	played	a	game
She	will	have	played	a	game
It	will	have	played	a	game
They	will	have	played	a	game

From the above example, it will be seen that the structure of sentences in the future perfect tense is as follows :

Subject + shall/will + have + Main Verb (past participle) + Object.

Usage : The future perfect tense is used to express an action that would have just completed at a point of time in the future.

4.	**The Future Perfect Continuous Tense :** The future perfect continuous tense represents an action which might begin at some definite time in the future partly completed at the particular point of time in future and would be in progress thereafter also in future.

I	shall	have	been	writing	a	letter
We	shall	have	been	writing	a	letter
You	will	have	been	writing	a	letter
He/She/It	will	have	been	writing	a	letter
You	will	have	been	writing	a	letter
They	will	have	been	writing	a	letter

From the above examples it will be seen that the structure of sentences in the future perfect continuous tense is as follows :

Subject + shall/will + have + been + Main Verb (in the present participle form i.e. verb + ing)
ι Object.

Usages : The future perfect continuous tense is used to express an action which will begin at some definite time in future shall be partly completed at, a particular time and shall be in progress for some time in the future beyond the particular time.

A Chart showing the various tenses [Active Voice + Form]   **M.V. = Main Verb**

Simple Present	Present Continuous	Present Perfect	Present Perfect Continuous
Subject + M.V. + Object I play a game. We play a game. You play a game. He/She/It plays a game. They play a game.	**Subject + is/am/are + M.V. (ing) + Object** I am playing a game. We are playing a game. You are playing a game. He/She/It is playing a game. They are playing a game.	**Subject + has/have + M.V. (Past Participle + Object** I have played a game. We have played a game. You have played a game. He/She/It has played a game. They have played a game.	**Subject + has/have + been + M.V. (ing) + Object.** I have been playing a game. We have been playing a game. You have been playing a game. He/She/It has been playing a game. They have been playing a game.
Simple Past	**Past Continuous**	**Past Perfect**	**Past Perfect Continuous**
Subject + M.V. (in paste tense) + Object I played a game. We played a game. You played a game. He/She/It played a game. They played a game.	**Subject + was/were + M.V. (ing) + Object** I was playing a game. We were playing a game. You were playing a game. He/She/It was playing a game. They were playing a game.	**Subject + had + M.V. (Past Participle) + Object** I had played a game. We had played a game. You had played a game. He/She/It had played a game. They had played a game.	**Subject + had + been + M.V. (ing) + Object.** I had been playing a game. We had been playing a game. You had been playing a game. He/She/It had been playing a game. They had been playing a game.
Simple Future	**Future Continuous**	**Future Perfect**	**Future Perfect Continuous**
Subject + shall/will + M.V. + Object I shall play a game. We shall play a game. You will play a game. He/She/It will play a game. They will play a game.	**Subject + shall/will + M.V. (ing) + Object** I shall be playing a game. We shall be playing a game. You will be playing a game. He/She/It will be playing a game. They will be playing a game.	**Subject + shall/will + have + M.V. (Past Participle) + Object** I shall have played a game. We shall have played a game. You will have played a game. He/She/It will have played a game. They will have played a game.	**Subject + shall/will + have + been + M.V. (ing) + Object.** I shall have been playing a game. We shall have been playing a game. You will have been playing a game. He/She/It will have been playing a game. They will have been playing a game.

A chart showing various Tenses Passive Form $\begin{bmatrix} \text{M.V.} = \text{Main Verb} \\ \text{P.P.} = \text{Past Participle Form of Main Verb} \end{bmatrix}$

Tense	Indefinite or Simple	Continuous or Imperfect	Perfect	Perfect Continuous
Present	Subject is/am/are + M.V./P.P. + by + object I am liked We are liked You are liked He/she/it is liked They are liked	Subject is/am/are + being + M.V./P.P. + by + object I am being liked We are being liked You are being liked He/she/it is being liked They are being liked	Subject have/has + been + M.V./P.P. + by + object I have been liked He/she/it has been liked You have been liked They have been liked	Not in use
Past	Subject was/were + M.V./P.P. + by + object I was liked We were liked You were liked He/she/it was liked They were liked	Subject was/were + being + M.V./P.P. + by + object I was being liked We were being liked You were being liked He/she/it was being liked They were being liked	Subject had + been + M.V./P.P. + by + object I had been liked We had been liked You had been liked He/she/it had been liked	Not in use
Future	Subject shall/will + be + M.V./P.P. + by + object I shall be liked We shall be liked You will be liked He/she/it will be liked They will be liked	Not in use	Subject shall/will + have + been + M.V./P.P. + by + object I shall have been liked We shall have been liked You will have been liked He/she will have been liked It will have been liked They will have been liked	Not in use

Fill in the blanks with the appropriate form of the verbs given in the brackets. Answers to the First twenty sentences have been given in a bracket against each.

1. He (go) to polytechnic every day. (**goes**)
2. The sun (rise) in the east. (**rises**)
3. I (sit) on a chair and (eat) an apple. (**sat, ate**)
4. Some students never (study) hard. (**study**)
5. The teacher (point) at the black board, when he,(want) to (write) something. (**points, wants, write**)
6. Engineers (make) the plans of the buildings. (**make**).
7. He always (meet) her in the institute. (**meets**)
8. The baby (cry) because of pains. (**cried**)
9. Aeroplane (fly) from Bombay to Delhi everyday. (**flies**)
10. He (travel) to Bombay tomorrow. (**will travel**)
11. See, a man (run)-after the bus. He (want) to get in. (**runs, wants**)
12. See, how, he (run). (**runs**)
13. The sun (warm) the air and (give) us light. (**warms, gives**)
14. Seeta (study) for an examination now. (**is studying**)
15. Wood (float) on water but iron not float. (**floats, does not float**)
16. He (read) when-she came in. (**was reading**)
17. The sun (shine) when he went out. (**was shining**)
18. He (sit) in the garden when the snake (come) out. (**was sitting, came**).
19. When they (came) in, he (write). (**came, was writing**)
20. I came in while he (sleep). (**was sleeping**)

<table>
<tr><td colspan="2" align="center">EXERCISE</td></tr>
<tr><td>(1)</td><td>It (rain) this morning when I got up.</td></tr>
<tr><td>(2)</td><td>We (work) all day yesterday.</td></tr>
<tr><td>(3)</td><td>We (live) in Poona when the war began.</td></tr>
<tr><td>(4)</td><td>The boy jumped off the bus while it (move).</td></tr>
<tr><td>(5)</td><td>The train started while I (get) on.</td></tr>
<tr><td>(6)</td><td>He (sit) in a hotel, when I (see) him.</td></tr>
<tr><td>(7)</td><td>When I (go) out the sun (shine).</td></tr>
<tr><td>(8)</td><td>The boy (fall) down while he (run).</td></tr>
<tr><td>(9)</td><td>The light (go) out while I (have) tea.</td></tr>
<tr><td>(10)</td><td>When it (rain) she (carry) an umbrella.</td></tr>
<tr><td>(11)</td><td>She (die) while she (run) after a bus.</td></tr>
<tr><td>(12)</td><td>When the phone (ring) I (have) a bath.</td></tr>
<tr><td>(13)</td><td>He waited for his friends until they (come).</td></tr>
<tr><td>(14)</td><td>He speaks as one who (know).</td></tr>
<tr><td>(15)</td><td>He (run) as fast as he could.</td></tr>
<tr><td>(16)</td><td>Her health improved since she (leave) Pune.</td></tr>
<tr><td>(17)</td><td>Whenever they (meet), they talk of old times.</td></tr>
<tr><td>(18)</td><td>He ran because he (to be) in a hurry.</td></tr>
<tr><td>(19)</td><td>He eats as much as he (can).</td></tr>
<tr><td>(20)</td><td>She rode as swiftly as she (can).</td></tr>
<tr><td>(21)</td><td>He ran away because he (to be) afraid of the danger.</td></tr>
</table>

(22) I walked so fast that he (can) not overtake me.
(23) When I (see) him, he (sit asleep) in the chair.,
(24) He (write) a letter now.
(25) He already (write) two letters this morning.
(26) I (learn) English for the last several years and now I (study) Hindi too.
(27) My wife (not come) home yet. She never (come) home before night.
(28) When water (boil) the liquid (change) to vapour that (be call) steam.
(29) She never (see) the sea. She (want) to go last year but she (have) no money.
(30) I (expect) you (hear) how he (win) a medal for bravery.
(31) Last year they (begin) to build a new building.
(32) The town (change) its appearance since 1960.
(33) Two years ago they (call) in an expert architect.
(34) He already (design) some public buildings.
(35) Sometimes the roofs (leak) and (let) in the rain.
(36) They already (repair) some of the machines and (make) them more suitable for work.
(37) Lend me your pencil. I (have) to write a letter.
(38) I (see) you yesterday. You (drink) tea in the hotel, but you (not see) me.
(39) Butter is usually (pack) in packets of 100 gms.
(40) English (study) almost everywhere.
(41) His book (read) everywhere.
(42) Many things (make) from plastic.
(43) Electricity (generate) in a power station.
(44) The news (broadcast) at 9.00 p.m.
(45) That scientist (teach) me a lot since 1970.
(46) If we set out late tomorrow, we (be) late for the function.
(47) While he (write) a letter, the telephone rang.
(48) He (go) to the college in time everyday.
(49) Next year by this time we (complete) our work.
(50) Aeroplanes only became possible when suitable engine (invent).
(51) I (wish) I (know) the answer now.
(52) He (look) very nervous when I (meet) him yesterday.
(53) I (lose) my English book yesterday.
(54) The patient (die) before the doctor came.
(55) I (hear) a knock just after I (finish) the story.

THE VERB : PERSON AND NUMBER

1. The verb has three persons viz., the first the second and the third.

   For example : (i) I talk (verb in first person)

   (ii) You talk (verb in second person)

   (iii) He talks (verb in third person)

   Verb takes the same person as its subject.

2. The verb, like the noun and pronoun, has two numbers viz. singular and plural.

   For example : (i) He talks (verb in singular)

   (ii) They talk (verb in plural).

   It means that the verb agrees with its subject in Number.

3. The verb must agree with its subject both in number and person.

     For example :   (i)   I am here
                            (ii)   They are here
                            (iii) He is here.

4. In modern English, verbs have lost their inflections except in the second and the third person of the singular number.

     For example :   I talk   We talk
                            You talk   You talk ('you' is both singular and plural)
                            He talks   They talk

The verb ' to be ' is an exception. It takes different forms with different persons.

     For example :   I am   We are
                            You are   You are
                            He/She/It is   They are

Agreement of the verb with the subject

1. The verb agrees with the subject in number and person

    (i) In case two or more subjects are connected by 'and', they usually take a verb in the plural.

        For example :   Ram and Hari **are** going out.
                              He and his friend **have** left.

    (ii) In case two singular nouns refer to the same person or thing, the verb takes a singular form.

        For example :

        (i)   **The** orator and statesman **has** left. (It means that the orator and statesman are one and the same person **as only one article is used**)

        (ii)   **The** orator and **the** statesman **have** left. (It means that orator and statesman are two different persons as two separate articles are used)

        (iii) In case two subjects together express one idea, the verb must be in a singular form.

        For example : Slow and steady **wins** the race.

        (iv) In case singular subjects are preceded by each or every, the verb is usually in singular form.

        For example :   **Every** boy and girl is ready.

       Every man and woman were present for the function.

       Every day and each hour brings certain difficulties.

2. Two or more singular subjects connected by **or, nor, either or, neither nor** take a verb in the singular form.

        For example :   No nook **or** corner was left.
                              Neither he nor I **was** present.
                              Either Ram or Hari **has** done that work.
                              Neither Ram nor his brother **was** present.

3. In certain cases when subjects joined by or, nor, are of different numbers, the required verb must be in plural form and the plural subject has to be placed next to the singular subject.

        For example :   Ram or his sisters **have** failed to do this job.

                              Neither the teacher nor the students were present.

                              Either the boy or his parents have raised the objection.

4. When subjects of different persons are joined by or, nor the verb agrees with the nearest person.

    For example :   **Neither** he **nor I have** committed mistake.

                 Neither you nor he **is** taking responsibility.

5. In case of subjects differing in number, or person or both, are connected by **and**, the verb must be in plural number.

    For example :   You and he are birds of the same feather

6. A collective noun takes a singular verb when the collection is taken as a whole.

    For example :   The Parliament has elected him.

                 The fleet has set sail.

                 A group of students was present in the class.

                 The mob has wrong tendency.

7. Certain nouns which are plural in form but singular in meaning, take singular verb.

    For example :   **The News is** good.

                 Physics is a branch of science.

8. Certain nouns which are singular in form but plural in meaning, take a plural verb.

    For example :   Three dozen **cost** one hundred rupees.

9. In case plural noun comes between a singular subject and its verb, it should agree with the real subject.

    For example :   **Each** of the brothers **is** intelligent.

                 Each of the students **was** rewarded.

                 Neither of them **was** tall.

                 A variety of objects **charms** us.

                 The quality of the products **was** not upto the mark.

                 If it **were** possible to get glimpses, we would have become happy.

10. Words joined to a singular subject by "with" "together with", "In addition to" or "As well as" are parenthetical and do not affect the number of the verb.

    For example :   **The chief**, with all his men, **was** present.

                 Ram, as well as his brothers, **deserves** credit.

                 Ram, and not you, **has** won the game.

THE INFINITE

All verbs in the indicative, subjective and imperative mood are Finite, because they are limited by the person and number and their subject.

For example : They always **find** mistake in my work.

In this sentence the verb 'find' has 'they' for its subject, Hence the verb 'find' is limited by person and number. Hence such verb is called as 'Finite Verb '

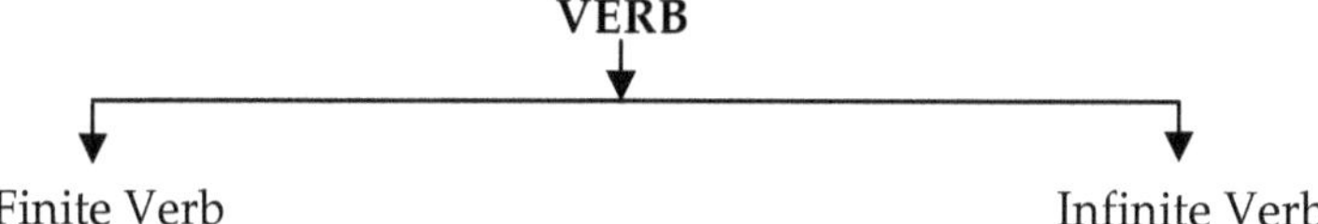

Infinite verb merely names the action denoted by the verb and is used without mentioning any subject. Hence, it is not limited by person and number as a verb that has a subject, and is, therefore, called the verb Infinite or simply that Infinitive.

For example : They always try **to find** mistakes in my work.

The Past Participle

A participle is that form of a verb which partakes of the nature both of a verb and of an object. (It means the verb is partly adjective and partly verb).

For example : Hearing the noise, he came out of the room.

In this sentence the phrase, 'Hearing the noise' which is introduced by a participle is called as 'participle phrase'. (It is an adjective phrase).

The Gerund

A Gerund is that form of the verb which ends with '– ing' and has the force of a noun and a verb.

For example : Reading is his favourite habit.

In the above sentence, the subject 'Reading' is a gerund. Like a noun, it is the subject of a verb, but like a verb, it also takes an object. It also shows the force of a verb. Such verbs are called gerund.

Strong and Weak Verbs

When a verb is changed into past tense by adding "-d", "ed " or 't' to its present tense form, it is called as a "Weak Verb".

For example :

Present Tense	Past Tense
Walk	Walked
Require	Required
Play	Played
Want	Wanted
Send	Sent

In case the verb changes into past tense merely by changing a vowel in its present tense form, it is called as a **strong verb**. In other words, if a verb does not require the addition of 'ed' or 'd' or 't' for changing it into the past tense form, it is called as a strong verb.

For example :

Present Tense	Past Tense
Catch	caught
Break	broke
Choose	chose
Teach	taught

A List of some useful strong verbs

Present Tense	Past Tense	Past Participle
Abide	abode	abode
Arise	arose	arisen
Bear (bring forth)	bore	borne
Bear (carry)	bore	borne
Beat	beat	beaten
Become	became	become
Beget	begot	begotten
Begin	began	begun

Present Tense	**Past Tense**	**Past Participle**
Behold	Beheld	beheld, beholden
Bid	bade, bid	bidded, bid
Bind	bound	bound
Bite	bit	bitten, bit
Blow	blew	blown
Break	broke	broken
Chide	chided	chided, chidden, chid
Choose	chose	chosen
Cling	clung	clung
Come	came	come
Dig	dug	dug
Do	did	done
Draw	drew	drawn
Drink	drank	drunk, drunken
Drive	drove	driven
Eat	ate	eaten
Fall	fell	fallen
Fight	fought	fought
Find	found	found
Fling	flung	flung
Fly	flew	flown
Forbear	forbore	forborne
Forbid	forbade	forbidden
Forget	forgot	forgotten
Forsake	forsook	forsaken
Freeze	froze	frozen
Get	got	got, gotten
Give	gave	given
Go	went	gone
Grind	ground	ground
Grow	grew	grown
Hide	hid	hid, hidden
Hold	held	held
Know	knew	known
Lie	lay	lain
Ride	rode	ridden
Ring	rang	rung
Rise	rose	risen
Run	ran	run
See	saw	seen
Shake	shook	shaken
Shine	shone	shone
Shoot	shot	shot

Present Tense	Past Tense	Past Participle
Shrink	shrank	shrunk, shrunken
Sing	sang	sung
Sink	sank	sunk
Sit	sat	sat
Slay	slew	slain
Slide	slid	slid
Sling	slung	slung
Slink	slunk	slunk
Smite	smote	smitten
Speak	spoke	spoken
Spin	spun	spun
Spring	sprang	sprung
Stand	stood	stood
Steal	stole	stolen
Stick	stuck	stuck
Sting	stung	stung
Stink	stank	stunk
Stride	strode	stridden
Strike	struck	struck
String	strung	strung
Strive	strove	striven
Swear	swore	sworn
Swim	swam	swum
Swing	swung	swung
Take	took	taken
Tear	tore	torn
Throw	threw	thrown
Tread	trod	trodden, trod
Wear	wore	worn
Weave	wove	woven
Win	won	won
Wind	wound	wound
Wring	wrung	wrung
Write	wrote	written

Exercise 1 :

Verbs

Use the following verbs in sentences of your own.

abominate	enrich	impede	snatch
aggravates	Facilitate	Inflicted	spill
Alleviate	fantasize	leap	stonewall
cease	flown	oxidize	surpass
dazzle	goaded	regulate	utilize
decipher	graft	rescue	venture
disrupt	ignited	segregate	yearn

Adverbs

A word which modifies the meaning of a verb, an adjective, or another adverb is called as an Adverb.

For example :   1.   He runs **quickly**.
                2.   He runs **more quickly**.
                3.   He is a **highly intelligent** person.

In the first sentence, the word 'quickly' shows the manner of running. Therefore, the word, quickly modifies or adds to the meaning of the verb 'run'. In the second sentence the word 'more' modifies the adverb 'quickly'. In the third sentence, 'highly' modifies an adjective 'intelligent'. All these three words viz. quickly, more and highly are the adverbs.

Kinds of Adverbs

1.   Adverbs of Time (before, daily, already etc.). (It shows when).
2.   Adverbs of Frequency (twice, again, once etc.). (It shows how often.)
3.   Adverbs of Place (here, up, out etc. (It shows where.)
4.   Adverbs of Manner (bravely, soundly, well etc.). (It shows how.)
5.   Adverbs of Degree or quality (almost, fully, quite etc.). (It shows how much, in what degree.)
6.   Adverbs of Affirmation and Negation (surely, certainly, not etc.).
7.   Adverbs of Reason (hence, therefore etc.).
8.   Simple adverbs (yes, no, when these are used as equivalent of sentence)

Position of Adverbs

1.   Adverbs of manner are generally used after the verb or after the object.
     For example :   The car is moving **slowly**.
                     He speaks **well**.
2.   Adverbs of place and of time are placed after the verb or after the object if there is one.
     For example :   **They** will come **here**.
                     He looked **everywhere**.
3.   When there are two or more adverbs after a verb (and its object), the normal order is, adverb of manner, adverb of place, adverb of time.
     For example :   He played **well in the games**.
                     He should **go there tomorrow morning**.
                     He spoke **sincerely at the meeting yesterday**.
4.   Adverbs of frequency (always, never, often, rarely, usually, generally, almost, nearly etc.) are normally put between the subject and the verb, if it consists of only one word. In case there are more than one word in the verb, they are placed after the first word.
     For example :   We **usually** take our meals at ten.
                     He has **never** observed such happening.
                     We **quite** agree with him.
5.   In case the verb is a form of 'to be' (am/are/is/was) an adverb is placed after the verb.
     For example :   I was **never** late for college.
                     They are **always** at home at night.
6.   Adverbs are usually placed before an auxiliary or the single verb 'be', when it is used for giving stress on the action.
     For example :   1.   "When will you complete this work ?" "But I **already** have completed it."
                     2.   " Do you work ? " "Yes, **I sometime** do ".

7. The auxiliaries 'have to' and 'used to' prefer the adverb before them.

    For example :   We **often** have to go to polytechnic on bicycle.

    He **always** used to agree with him.

8. An adverb that modifies an adjective or another adverb, usually comes before the word it modifies.

    For example :   He is a **rather** lazy student.

    The speech is **very** interesting.

9. The adverb 'enough' is always placed after the word it modifies.

    For example :   He was rash **enough** to stop.

    They spoke loud **enough** to be heard.

10. The word 'only' should be placed immediately before the word it modifies.

    For example :   He worked only two hours.

    They have taken only two hours to complete the work.

EXERCISE 1

Rewrite the following sentences appropriately using the words : out, off, through, by :

1. He has ruled ______ the division of the state.
2. He drag ______ hearings to enhance the fees.
3. The tension has eased ______ .
4. The plane will take ______ soon.
5. Our plan for holiday fell ______ because of his objection.
6. He will surely back ______ at the last moment.
7. Defective machinery wear ______ soon.
8. It can take ______ vertically.
9. It will put ______ the fire in short period.
10. Come ______ with concrete proposals.

EXERCISE 2

Rewrite the following sentences by appropriately completing the verb   phrases :

1. He laid ______ his life for his country.
2. His interest in studies has fallen ______ recently.
3. I was held ______ in traffic jam.
4. ______ keep cost in high in this plan.
5. He has not turned ______ at the meeting.
6. Speak ______ please.
7. I soon put ______ the light and lay ______ to sleep.
8. What do these letters stand ______ ?
9. Write it ______ in full, please.

EXERCISE 3

Use the following Adverbs in sentences of your own.

commendably	eventually	logically	quickly
Doggedly	Exclusively	manually	seemingly
easily	globally	obviously	slowly
Economically	ideally	physically	triumphantly
entirely	Intuitively	Predominantly	Utterly

3. DO AS DIRECTED

(Transformation of Sentences)

Sentences are units made up of one or more clauses. Every sentence has two parts :

1.   The part called **subject** which names the person or thing we are speaking about ;

2.   The part called **predicate** which tells something about the subject of the sentence.

The subject of a sentence usually comes first; but sometimes it can be used after the predicate. In case of imperative sentences, the subject is left out. See the following sentences :

1. | The earth | | is round | .

2. | The news | | made him happy | .

3. | They | | elected him President | .

We see that the above sentences have subject and predicate. Such sentences are called as simple sentences.

Sentences are classified on the basis of two aspects viz. Syntactic i.e. based on the arrangement of words, phrases and clauses and semantic i.e., based on the sense the sentence carries.

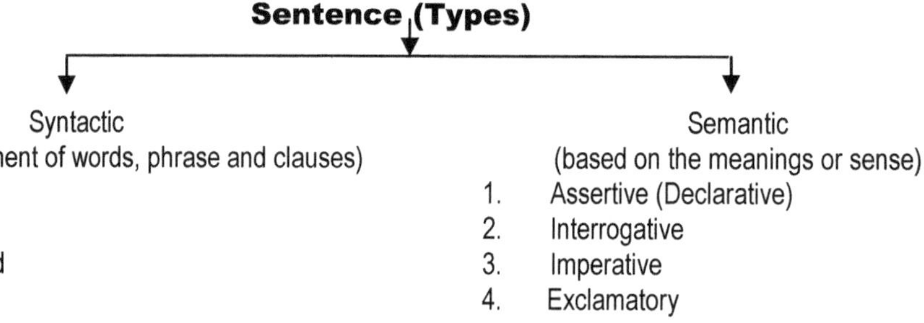

Sentence (Types)

Syntactic	Semantic
(based on arrangement of words, phrase and clauses)	(based on the meanings or sense)
1. Simple	1. Assertive (Declarative)
2. Complex	2. Interrogative
3. Compound	3. Imperative
4. Multiple	4. Exclamatory

Syntactic Types :

1.   **Simple sentences :** A simple sentence is one which has **one subject, one predicate and one finite verb.**

      For example : | The earth | | is round |

               subject   predicate
                               Finite verb

2.   **Complex sentence :** Complex sentence is one which consists of a main clause together with one or more subordinate clauses.

      For example : | He hoped | | that he would pass |

              main clause   subordinate clause

3.   **Compound sentence :** Compound sentence is one which contains two or more main clauses. These clauses are said to be co-ordinate, that is, of equal rank. Such clauses are joined by co-ordinate conjunctions such as and, but, or, nor, for, etc.

      For example : | God made the country | and | man made the town. |

            main clause   conjunction   main clause

4.   **Multiple sentence :** A multiple sentence is one which contains more than two main clauses and may contain one or more subordinate clauses and depending upon one of the main clauses.

      For example :   His mother said that he had gone to the market to make a few purchases but that I should not on that account go away as he would return soon.

Analysis of complex, compound and multiple sentences :

Analysis of a complex sentence involves detailed consideration and classification of the subordinate clauses. On the basis of the function they perform, the clauses are classified as noun clauses, adjective clauses and adverb clauses. They are briefly explained below :

The following are the types of subordinate clauses and the different ways they are used in the complex, compound and multiple sentences. The clauses are classified on the basis of the function they perform in a given sentence.

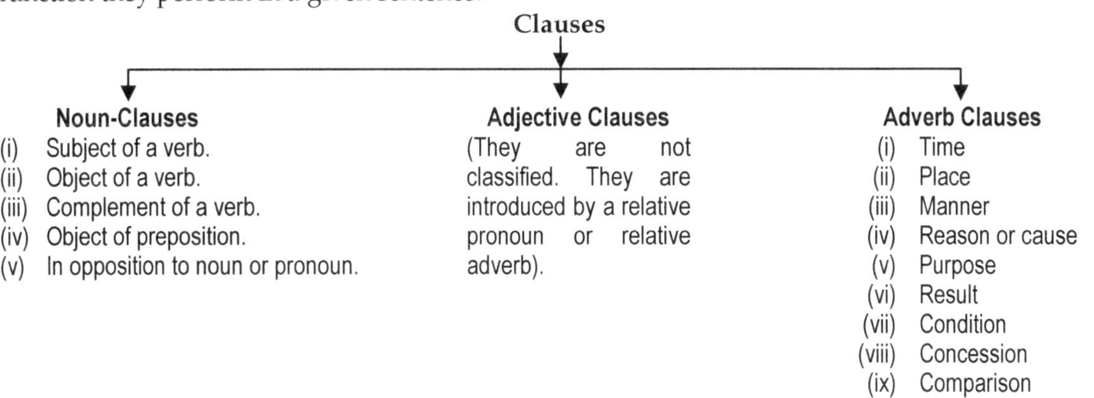

Noun clause : Noun clause can be used in five ways :

1. As a subject of a verb
2. As an object of a verb
3. Predicatively (complement)
4. As an object of a preposition
5. In apposition to a noun or pronoun.

1. **Subject of a verb**

   For example :   **That the distinguished visitor was about to enter the class** became apparent to the curious students.

2. **Object of a verb :**

   For example :   Ram asked the teacher **whether he was to be given another chance to answer the question.**

3. **Complement of a verb of incomplete predication :** Ram's excuse was **that he was inattentive.**

4. **Object of preposition :**

   For example :   **From what I heard** there will be no game today.

5. **In apposition to a noun or pronoun**

   For example :   1.   I am firm in my conviction **that he will win.**

                     2.   It is evident **that he has lost the game.**

Adjective clause : There is no 'kind' or 'classification' in adjective clause. Adjective clauses are mostly introduced by a relative pronoun or relative adverb. The relative pronoun or relative adverb is often omitted.

       For example :   1.   That is not the book (which) **I read last week.**

                       2.   The reason (why) **he left so hurriedly** is not known.

Adverb clause : There are nine kinds of adverb clauses showing time, place, manner, cause or reason, purpose, result, condition, concession, comparison or degree.

   1. Adverb clause of Time : An Adverb clause of Time is generally introduced by the following subordinate conjunctions :

      1.   Since   2.   When   3.   Whenever   4.   While

      5.   Until   6.   Till   7.   Before   8.   After.

For example : He entered the room | when the door opened |

↓

Adverb clause of Time.

2. **Adverb clause of Place**

    For example : Please accompany | where he goes |

3. **Adverb clause of Manner**

    For example : He played | as he was expected to play |

4. **Adverb clause of reason or cause**

    For example :   1. I returned home | because I could not see him |

                    2. | As he was ill | he could not attend the class.

5. **Adverb clause of purpose**

    For example :   1. He ran fast that | he might catch the bus |

                    2. He ran fast | lest he should miss the bus |

6. **Adverb clause of Result :** This clause is introduced by 'that' which is generally preceded by 'so' or 'such' in the principal clause.

    For example : She sang **such** sweet songs | that we were very happy |

7. **Adverb clause of condition :** This clause is introduced by the conjunctions like "If" and "unless". Sometimes, 'if' is not expressed but understood.

    For example :   1. | If you try | , you can succeed.

                 2. | Unless you are confident of success | , you should not make that venture.

                 3. | Had I been present | , I would have prevented the mishap.

8. **Adverb clause of concession :**

    For example : | Although he is poor | , he is honest.

9. **Adverb clause of comparison**

    For example :   (Introduced by 'than')

                    He played better | than his friend did | .

                    Sometimes a clause of comparison is suppressed after 'than'.

    For example :   He looks better than when I saw him last (than he did then)

Semantic Type

A group of words which makes complete sense is called a sentence. There are four kinds of sentences.

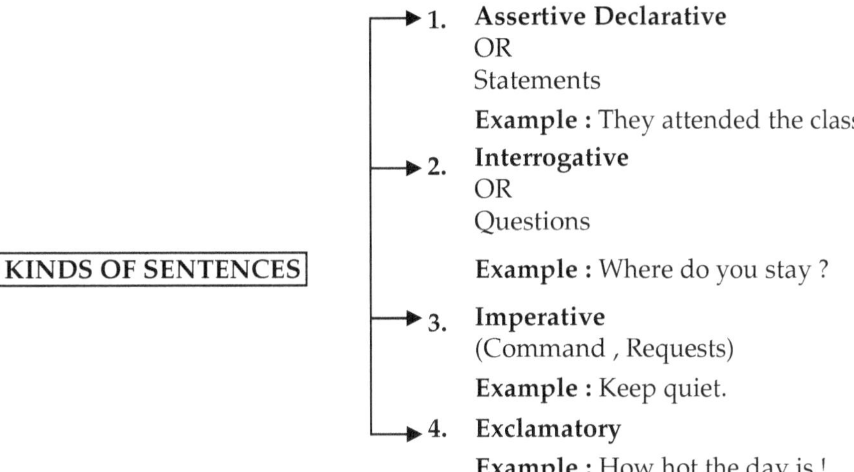

1. **A sentence which makes a statement or assertion is called as Declarative or Assertive sentence.**

   For example :   I have a good memory

                     The earth moves around the Sun

                     The Sun rises in the east

                     No one can serve two masters.

2. **Interrogative sentence :** Interrogatives are the words which introduce wh – questions. The English interrogative words are – who, whom, where, what, which whose, whether, why, when, how, if (They are taken as **Wh–words**). **Whether** and **if** are used only in interrogative Sub-clauses.

   For example :   What time is it ? (Determiner)

                     What's the time ? (Pronoun)

3. **Imperative (Command, Request)**

   A sentence that expresses a command is called as **Imperative sentence.** The subject of a verb in imperative sentence is usually omitted.

   For example :   Write properly.

                     Keep silence.

                     Go away.

                     Pay the bill.

   Exclamatory : A sentence that expresses strong feeling or emotion is called as an **Exclamatory sentence.**

   For example :   What a pity !

                     What a piece of work this is !

                     On that he is safe !

                     What a beautiful scene this is !

   It is this type of sentence, which is used to express the speaker's feeling or attitude.

(A) Active and Passive Voice

[Refer Chart given on page no. 2.20]

Voice is that form of the verb which shows whether the person (doer of the action) or thing denoted by the subject acts, or is acted upon.

A verb is said to be in the Active Voice when the person or thing denoted by the subject performs an action. An active verb cannot be turned into the passive, unless it has an object. In other words, only transitive verbs have a passive voice.

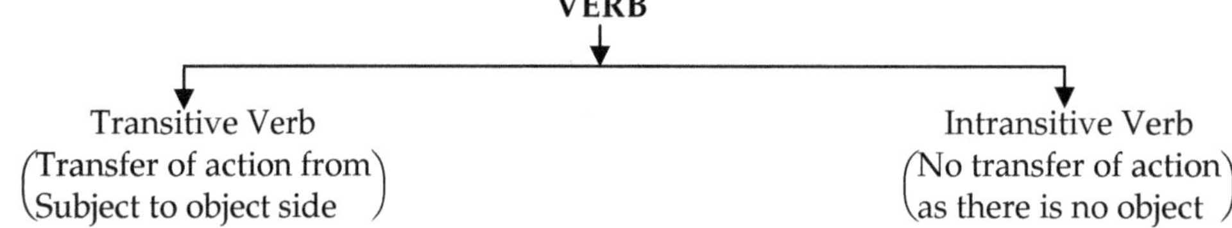

Example : I Wrote a letter.

(In this example, we can see that the verb 'wrote' is used transitively, as the sentence has an object, 'letter.' For performing that action, there is a transfer of action from the subject to the object. The above mentioned verb can, therefore, be said to be used transitively).

Example : He runs quickly.

In this example, the object is not used and hence there is no question of transfer of action. It can be said that the verb is used intransitively. Hence, such sentences cannot be changed into Passive Voice.

The following important points should be considered, while transforming a sentence from active voice to passive voice.

He	wrote	a letter
Subject	Verb	Object

I.   Mark out the subject(s), verb(v) and object in the given sentence ((o)) .

    (a) In some cases, when it becomes difficult to find the object from the given sentence, it is better to ask the question "What + S + V." The answer we get is the exact 'object.'

               What he wrote .... a letter (object)

    (b) The sentence must have an object or an object understood, without which we cannot transform the given sentence from active voice to passive voice.

II. Interchange the places of the words used at the place of subject and object. (Object becomes subject & subject becomes object)

He	wrote	a letter
Subject	Verb	Object
A letter :	was written = by	+ him
(Subject)	Verb	(Object)

Subjective Pronouns		**Objective Pronouns**
1.	I	me
2.	We	us
3.	You	you
4.	He	him
5.	She	her
6.	It	it
7.	They	them

III. Use one of the forms of the verb 'to be' according to tense and number of the subject taken in the Passive Voice.

(Simple Past)   He   wrote   a letter   (Active voice)
                     S   V   O

A letter [was] .... him

Active Voice Sentence		**Use one of the To be's form**
1. Present simple	→	is/am/are
2. Past simple	→	was/were
3. Future Simple	→	shall/will be
4. Continuous (Present/Past/Future)	→	being
5. Perfect (Present/Past/Future)	→	been
6. Perfect Continuous	→	(Such Passive Voice sentences are not in use)

IV. Use the main verb in the past participle form in Passive Voice.

(Simple Past)   He   wrote   a letter   (Active voice)
                     S   V   O

A letter was written by <u>him</u>
S   O
(Past Participle form)

V. Use 'by' as a preposition before the object.

(Simple Past)   He   wrote   a letter   (Active voice)
                     S   V   O

A letter was written by <u>him</u>.

Subject   by   Object
(Auxiliary (main verb   Preposition
verb To   in past
be's group Participle Form).

To sum up, let us see what we have done while converting the active voice sentence. 'He wrote a letter' into a passive voice sentence. A letter was written by him.'

Here,

(i) **Subject became object ;**

(ii) **Object became subject ; (subjective pronoun becomes objective pronoun)**

(iii) **Use appropriate form of the auxiliary verb 'to be' with reference to the subject and tense of the verb.**

(iv) **Main verb is converted into 'past participle' form.**

(v) **Use of preposition 'by' before the new object. (Refer Page No. 2.20)**

Now, we will try to transform the same sentence, taking it in all other possible tenses.

I. [Present simple] <u>He</u>   <u>writes</u>   a <u>letter</u> [Active voice]
                   S   V   O
A letter is written by him. [Passive voice]
[Structure in Subject + is/am/are + Main Verb in Past Participle form + by + Object
Passive voice

II. [Past simple]   <u>He</u>   <u>wrote</u>   a <u>letter</u>   [Active voice]
                     S   V   O
A letter was written by him.   [Passive voice]
[Structure in   Subject + was/were + Main Verb in Past Participle form + by + Object
Passive voice]

III.   [Future simple] He   will write   a Letter   [Active voice]
                                S   V   O

                 A letter will be written by him.   [Passive voice]

                               Verb
[Structure in Subject + shall/will + be + Main Verb in Past Participle form + by + Object Passive voice]

IV.   [Present   He   is writing a letter   [Active voice]
       continuous]   Subject   Verb   Object
                    A letter   is being written   by him   [Passive voice]
                    Subject   Verb   Object
       [Structure in   Subject + is/am/are + being + Main Verb in Past participle form + Passive voice] by + object.

V.   [Present   He   was writing   a letter   [Active voice]
                  A letter was being written by him. [Passive Voice]
       [Structure in Subject + was/were + being + Main verb in Past Participle form + by passive voice]   + Object.

VI.   [Future   He will be writing a letter   [Active voice]
       continuous]
                   (Passive voice form of sentences are not in use).

VII. [Present perfect]   He   has written   a letter.   [Active voice]
                        S   V   O
             A letter has been written by him. [Passive voice]
       [Structure in   Subject + has/have + been + Main verb in Past Participle form + Passive voice]   by + Object.

VIII.   [Past perfect]   He had written a letter. [Active voice]

             A letter had been written by him. [Passive voice].
       Structure in   Subject had + been + Main Verb in Past Participle Form + by Passive voice]   + object.

IX.   [Future Perfect Continuous]   He will have written a letter.
       [Active Voice]
                          A letter will have been written by him. [Passive Voice]
       [Structure in   Subject + shall/will + have + been + Main Verb in Past participle passive voice]   form + by + Object.

X.   Passive voice of all Present / Past / Future Perfect Continuous Tenses are not in use.

Imperative sentence : Pay   the bill   today
                         Main Verb   Object
           Let the bill be paid today. [Passive voice]
[Structure] : Let + Subject + be + Main verb in Past Participle Form + Remaining part of   the sentence.

     Imperative sentences start with the main verb and the subject always remains understood. Therefore, the Passive Voice Sentence starts with 'let' and the other part of the sentence structure follows the same method, which we have followed in case of the previous sentences.

Interrogative Sentences :

     Example 1 :   Who taught you this subject ?
                      By whom were you taught this subject ? [Passive voice]

{If 'who' appears in the beginning of an interrogative sentence, that changes into 'by whom').

The structure of the sentences changes as follows :

Interrogative Pronoun + Auxiliary Verb + Subject + Main Verb + Object + Question mark ?

Example 2 :	Does he write a letter ?

It is a letter written by him ?

(Here, in case the verb do and its form 'did' and 'does' are used in active voice, they are not taken ; because in the passive voice, we take one auxiliary verb i.e. to be with its form appropriate to subject and tense of the verb used.

Change the Voice :

1.	We can control the number of splitting atoms.

Ans.	The number of splitting atoms can be controlled by us.

2.	Engineers bore holes, every hundred yards.

Ans.	Holes are borne every hundred yards by engineers.

3.	They invited me to deliver a speech.

Ans.	I was invited by them to deliver a speech.

4.	The Committee has prepared the plan of the project in a great hurry.

Ans.	The plan of the project has been prepared in a great hurry by the Committee.

5.	We shall have to look into the matter.

Ans.	The matter will have to be looked into by us.

6.	We have to knock down the buildings.

Ans.	The buildings have to be knocked down by us.

7.	The directors might turn down our proposal to employ more assistants.

Ans.	Our proposal to employ more assistants might be turned down by the directors.

8.	He taught me Physics at the University.

Ans.	I was taught Physics at the University by him.

OR

Physics was taught to me at the University by him.

9.	The General Manager has promised the employees to more wages.

Ans.	The employees have been promised to more wages by the General Manager.

OR

More wages have been promised to be given to the employees by the General Manager.

10.	We shall send our sister a nice Diwali gift.

Ans.	A nice Diwali gift will be sent to our sister by us.

11.	People say that diffused light is good for reading.

Ans.	It is said by the people that diffused light is good for reading.

12.	The scientists believe that there is no life on Mars.

Ans.	It is believed by the scientists that there is no life on Mars.

13.	The artisans suggested that the raw materials should be of good quality.

Ans.	It was suggested by the artisans that the raw materials should be of a good quality.

14.	We shall recommend that the factory should have a computer.

Ans.	It will be recommended by us that the factory should have a computer.

15. It is time to call over the names.

Ans. It is time for the names to be called over.

16. I wrote a letter.

Ans. A letter was written by me.

17. He gave her the remaining money.

Ans. The remaining money was given to her by him.

OR

She was given the remaining money by him.

18. Take him to the King.

Ans. Let him be taken to the King.

19. Turn the ship homewards.

Ans. Let the ship be turned homewards.

20. Give me back my dear Eurydice.

Ans. Let my dear Eurydice be given back to me.

21. He never took it seriously.

Ans. It was never taken seriously by him.

22. Keep this in mind.

Ans. Let this be kept in mind.

23. Everyone knows her.

Ans. She is known by everyone.

24. One expects better behaviour from a college student.

Ans. Better behaviour is expected from a college student.

25. Who called your names ?

Ans. By whom were your names called ?

26. Who taught you grammar ?

Ans. By whom were you taught Grammar ? OR

By whom was Grammar taught to you ?

27. Did you write the letter ?

Ans. Was the letter written by you ?

28. What can we do now ?

Ans. What can be done by us now ?

29. How much do you earn ?

Ans. How much is earned by you ?

30. The incoming President is organizing a National Provincial Service.

Ans. A National Provincial Service is being organised by the incoming President.

EXERCISE

Change the voice in the following sentences :

1. The telephone wire has been cut.
2. Have you not finished the work yet ?
3. I declined the offer.
4. The firm has been supplying us inferior goods.
5. Who broke this jug ?

6. It is time to shut up the shop.
7. Tools cannot be made by animals.
8. They were questioning the thief.
9. We are all liable to forget this.
10. When you have read these, cast them into the fire.
11. She is selling her old ornaments.
12. Votes are cast by registered voters.
13. The criminal was arrested by the police.
14. Who drew that picture on the wall ?
15. Why were you punished by the teacher ?
16. Promises should be kept.
17. The fire damaged the building.
18. We offered her a chair.
19. Advertise the post.
20. The young man did not write home any letter.
21. Their army has been defeated by us.
22. Some one has picked my pocket.
23. You must endure what you cannot cure.
24. I shall explain this later.
25. The master appointed him monitor.

(B) Direct and Indirect Speech

In the English language, there are two ways of reporting conversation or a talk. They are known as :

1. Direct Speech and
2. Indirect Speech.

A proper understanding of these methods is essential in order to make efficient use of the English language.

1. **The Direct Speech :** It reports the exact words of the speaker, and they are enclosed with single or double quotation marks.

    For example : He said, "I am playing a game."

    In this example, the sentence or the words uttered by the speaker are reproduced by the reporter without making any changes. If the sentence is reproduced without changes or additions, it is called as a Direct Speech.

2. **Indirect Speech :** Indirect speech reports the words of the speaker, without quotation marks, i.e. in an indirect way. Here, the exact words of the speaker are not generally reproduced. Consider the following sentence :– He said that he was playing a game.

    This is an example of an indirect speech. In this sentence, the verb 'said' introduces the speech to be reported. It is called as a "Reporting verb" and the speech is called a "Reported Speech". The reporting verb is connected by the joining word 'that', after making a few changes in the original speech and after removing the quotation marks.

This will be more clear from the following illustration :

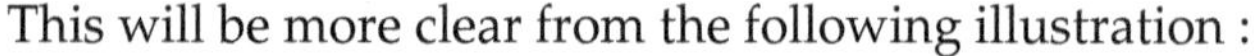
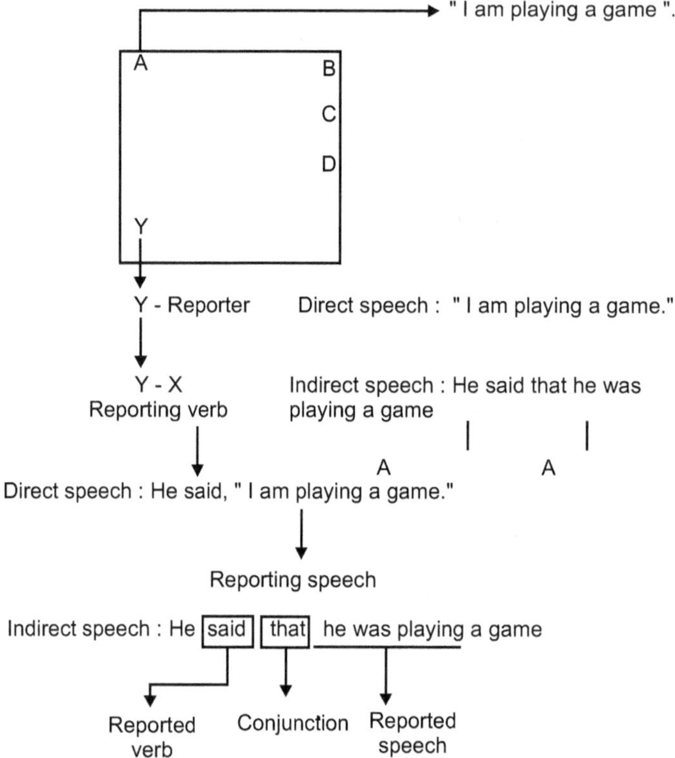

Necessary points to be remembered

1. Tenses   2. Reporting verb   3. Kinds of sentences   4. Pronouns,   5. Nearness.

Rules for the conversion of Direct Speech into Indirect Speech

1. The quotation marks should be removed.
2. A suitable joining word should be used to report the statements, particularly in the case of questions and exclamatory sentences.
3. Generally, the first and the second person pronouns in the direct speech are changed to the appropriate third person pronouns in the direct speech. The pronouns in the third person remain unchanged.
4. Whenever the antecedents of pronouns appear to be doubtful, it is better to put the antecedents within brackets after the pronouns.
5. Unnecessary modification in the structure of a sentence should be avoided.

   Change in Tenses : (In transformation from direct to indirect speech)

(a) If the reporting verb is in the present tense, the tense of the reported speech remains unchanged.

   For example :

Direct speech   :   He says, "I play a game."

Indirect Speech   :   He says that he plays a game.

                :   R.V. : Present Tense [R.V. – Reporting Verb]

Here, the reporting verb is in the present tense, hence the direct speech remains in the same tense i.e. present tense.

(b) If the reporting verb is in the future tense, the tense of the verb in the reported speech remains unchanged.

Direct   :   He will say, "I play a game."

Indirect :   He will say that he plays a game.

Here, the reporting verb is in the future tense, hence the Indirect Speech remains in the future tense.

(c) If the reporting verb is in the past tense, the verb in the reported speech must be changed to its corresponding form in the past tense.

Direct speech :

Present Simple	:	He said, "I play a game."
Present Continuous	:	He said, "I am playing a game."
Present Perfect	:	He said, "I have played a game."
Present Perfect Continuous	:	He said, "I have been playing a game."

Indirect Speech

Past Simple	:	He said that he played a game.
Past Continuous	:	He said that he was playing a game.
Past Perfect	:	He said that he had played a game.
Past Perfect Continuous	:	He said that he had been playing a game.

(d) If the reporting verb is in the past tense, the tense of the verb in the reported speech remains unchanged.

For example :

Direct speech   :   He said, "I played a game."

Indirect speech   :   He said that he played a game.

(Here the reporting verb is in the past tense, hence the reported speech is kept in the same tense).

Note : *Sometimes, the simple past tense of the reported speech changes into the past perfect tense.*

For example :

Direct speech   :   He said, "I played a game yesterday." (Simple past tense)

Indirect speech   :   He said that he had played a game the day before. (Past perfect tense)

Direct speech   :   He said, "I was playing a game." (Past continuous tense)

Indirect speech   :   He said that he was playing a game. (Past continuous tense)

Note : *Sometimes, the past continuous tense changes into the past perfect continuous if the specific point of time in the past tense is indicated in the direct speech.*

For example :

Direct speech   :   He said, "I was playing a game yesterday."

                 :   (Past continuous tense with specific time i.e. yesterday)

Indirect speech   :   He said that he had been playing a game the day before.

                 :   (Past perfect continuous)

Direct speech   :   He said, "I had played a game." (Past perfect tense)

Indirect speech   :   He said that he had played a game.

                 :   (Past perfect tense)

                 (Here, the tense remains unchanged).

Direct speech   :   He said, "I had been playing a game."
Indirect speech   :   He said that he had been playing a game.
                        (The tense in the reported speech also remains unchanged).

(e) If the reporting verb is in the past tense and the reporting speech is also in the past tense, the indirect speech remains unchanged, but sometimes, it changes as shown below :

Direct speech	Indirect speech

1.   Past Simple
    1. He said, "I played a game."
    Sometimes :
    2. He said, "I played a game yester-day."
	1. Past simple :
	    He said that he played a game. (No change in tense).
	    Sometimes : He said that he had played a game the day before. (Past perfect)

2.   Past continuous :
    1. He said, "I was playing a game."
    Sometimes :
    2. He said, "I was playing a game yesterday."
	2. Past continuous : He said that he was playing a game. (No change in tense)
	    Sometimes : He said that he had been playing a game the day before (Past perfect tense).

3. Past Perfect : He said, "I had played a game."	3. Past Perfect : He said that he had played a game. (No change in tense)

4.   Past Perfect :
    He said, "I had been playing a game."
    Continuous.
	4. Past Perfect Continuous : He said that he had been playing a game. (No change in tense)

If the reporting verbs are in the past tense and the tense of the reporting speech is in the future tense, the reported speech changes as follows :

Direct speech		Indirect speech
1. Future simple	: He said, "I will play a game."	1. He said that he would play a game.
2. Future continuous	: He said,"I will be playing a game."	2. He said that he would be playing a game.
3. Future Perfect	: He said, "I will have played a game."	3. He said that he would have played a game.
4. Future Continuous	: He said, "I will have been playing a game."	4. He said that he would have been playing a game.

Note : *If the reporting verb is in the past tense, the tense of the auxiliary verb in the reported speech is changed into the corresponding past tense.*

For Example :

1.   Can changes to could.
2.   May changes to might.
3.   Will changes to would.
4.   Shall changes to should.

Changes in the Pronouns in Reported Speech

1. The first person pronoun is changed to the corresponding third person pronoun in the reported speech, if the speaker is in the third person.

For example :

Direct speech   :   He said, "I play a game".
Indirect speech   :   He said that he played a game.
Direct speech   :   He said, "I – we – me us"
Indirect speech   :   He said that he they him them

2. If the speaker is in the second person, the first person pronoun in the reported speech is changed to the second person.

   For example :
   Direct speech   :   You said, "I play a game."
   Indirect speech   :   You said that you played a game.

3. If the speaker is in the first person, the first person pronoun in the reported speech is not changed.
   Direct speech   :   I said, "I am writing a letter."
   Indirect speech   :   I said that I was writing a letter.

4. A second person pronoun, in direct speech is changed in the indirect speech, to the same person of the pronoun being the object of the reported verb or the preposition following the reported verb.

   For example :
   1. Direct speech   :   He said to him, "You play a game."
      Indirect speech   :   He told him that he played a game.
   2. Direct speech   :   He said to you, "You play a game."
      Indirect speech   :   He told you that you played a game.
   3. Direct speech   :   He said to me, "You play a game."
      Indirect speech   :   He told me that I played a game.
   4. A third person in the direct speech remains unchanged in the indirect speech.
      Direct speech   :   He said, "He plays a game."
      Indirect speech   :   He said that he played a game.
      Direct speech   :   He said, "He, she, it – they him, her, it, them

   Indirect speech   :   He said that he, she, it they him, her, it them

Adjectives and adverbs indicating nearness in the direct speech are changed to the corresponding adjectives and adverbs indicating distance in the indirect speech as shown below :

	Direct speech	**Indirect speech**
1.	Now	then
2.	this	that
3.	these	those
4.	here	there
5.	ago	before
6.	thus	so
7.	today	that day
8.	tomorrow	the next day
9.	yesterday	the day before
10.	last night	the night before.

Interrogative sentences :

If an interrogative sentence is to be converted into indirect speech, it is necessary to make the following changes :

1. For introducing the reported speech, instead of the verb 'said', the verbs like 'asked', 'questioned', 'enquired' etc. may be used.

2. The joining word 'that' is not used to introduce the reported speech.

3. If the question begins with interrogative pronouns or interrogative adverbs such as which, when, whom, whose, where, why, how, etc. the same is retained to introduce the reported speech.

4. If the question begins with an auxiliary verbs like shall, will, do, etc., with forms of to be, the conjunction 'if' or 'whether' is used to introduce the reported speech.

5. The interrogative form of a sentence is changed to the statement form.

6. Interrogative mark '?' is not required in indirect speech.

Example No. 1 :

Direct speech   :   He said to him, "Where are you going ?"

Indirect speech   :   He asked him where he was going.

In this sentence, the reporting verb 'said' is changed into asked and the interrogative form of the reporting speech is changed into a statement form, by changing the places of the subject and verb of the Reported Speech).

Example No. 2 :

Direct speech   :   He said to me, "What are you doing ?"

Indirect speech   :   He asked me what I was doing.

Example No. 3 :

Direct speech   :   The teacher said to her, "Why do you laugh ?"

Indirect speech   :   The teacher asked her why she laughed.

In this direct sentence, 'do' is used to make the sentence interrogative; but in Indirect Speech, the sentence is changed into a statement removing the auxiliary verb 'do'.

Example No. 4 :

Direct speech   :   He said to me, "Are you not well today ?"

Indirect speech   :   He asked me whether I was not well that day.

If the direct speech starts with an auxiliary verb and it is an interrogative sentence, the conjunction 'whether' or 'if' is used to introduce the indirect speech.

Example No. 5 :

Direct speech   :   The boy said to father, "May I go out?"

Indirect speech   :   The boy asked the father if he might go out.

Imperative Sentences

In imperative sentences, we come across orders, commands and requests. The sense of the direct speech determines the rule.

Example 1 :

Direct speech   :   He said to the players, "Leave the ground at once."

Indirect speech   :   He ordered the players to leave the ground at once.

The following method should be adopted for transforming an imperative sentence in the direct speech into indirect speech.

1. Rewrite the part which comes before the Reporting verb.

2. The reporting verb 'said' is to be changed to 'ordered, informed, instructed, directed, commanded' or any other word with the same meaning and used after the subject.

3. The preposition 'to' that appears after 'said', should be removed.

4.   The part which appears after 'to' should be written without making any changes.

5.   As the Imperative form changes into the infinitive form, 'to' should be used and the complete sentence without any change is to be rewritten after making suitable changes (according to the rule) in the pronoun.

Example 2 :

Direct speech :	He	said to	his father,	"Pardon	me	this	time and I shall do my best."
Indirect speech :	He	begged	his father to	pardon	him	that	time and he would do his best.

Example 3 :

Direct speech :	He	said to	me,	"Let us go home".
Indirect speech :	He	proposed	me	that we should go home.

Example 4 :

Direct speech   :   My mother said to me, "Never make that mistake."

Indirect speech   :   My mother warned me never to make that mistake.

Exclamatory Sentences

In this type of sentence, the exclamatory words and marks of exclamations are used.

For example,

Direct speech   :   They said, "Hurrah! We have won the game."

Indirect speech   :   They exclaimed with joy that they had won the game.

Direct speech   :   They said, "Alas!. We are undone."

Indirect speech   :   They exclaimed with sorrow that they were undone.

In the case of exclamatory sentences, the following changes are to be made while changing the direct speech into indirect speech.

1.   The word of exclamation should be suitably changed into indirect speech and should be supported by an appropriate word indicating the underlying feelings such as sorrow, joy, etc. in the exclamatory verbs.

2.   Here, the conjunction 'that' is used.

3.   Other changes in pronouns and tense are done as in the case of assertive sentences.

Some more Examples :

1.   Direct Speech : He said, "I am ready to go out,"

     Indirect Speech : He said that he was ready to go out.

2.   Direct Speech : I said to him, "My sister has gone out."

     Indirect Speech : I told him that my sister had gone out.

3.   Direct Speech : Hari said to me, "I have been reading for four hours."

     Indirect Speech : Hari told me that he had been reading for four hours.

4.   Direct Speech : She said to her mother, "Sister was writing a letter."

     Indirect Speech : She told her mother that her sister was writing a letter.

5.   Direct Speech : The teacher said to him, " You may come."

     Indirect Speech : The teacher told him that he might come.

6. Direct Speech : The teacher said, "Man <u>is</u> mortal."

    Indirect Speech : The teacher said that man <u>is</u> mortal.

7. Direct Speech : He said to me, "When do you intend to leave Pune ?"

    Indirect Speech : He asked me when I intended to leave Pune.

8. Direct Speech : Hari said to me, "Thank you for the help." I could not have completed the task without your help".

    Indirect Speech : Hari thanked me for all my help and said that he could not have completed the task without my help.

9. Direct Speech : The teacher said to the boy, "Have you completed your writing work ?"

    Indirect Speech : The teacher asked the boy whether he had completed his writing work.

10. Direct Speech : The teacher said to the boy, "Go out of the room at once."

    Indirect Speech : The teacher ordered the boy to go out of the room at once.

11. Direct Speech : Hari said to me, "Let us play a game."

    Indirect Speech : Hari proposed playing a game with me.

12. Direct Speech : He said, "The bird died in the night"

    Indirect Speech : He said that the bird had died in the night.

13. Direct Speech : My teacher often says to me, "If you do not study hard, you will fail."

    Indirect Speech : My teacher often warns me that if I do not study hard, I will fail.

14. Direct Speech : The Captain, addressing his troops, said, "You have brought disgrace upon a famous regiment. If you had grievances, why did you not lay them before your own officers ? Now, you must first suffer punishment for your offence before your complaints can be heard."

    Indirect Speech : The Captain told his troops that they had brought disgrace upon a famous regiment. He asked them why they had not laid their grievances before their own officers. He then stated that they must suffer punishment for their offence, before their complaints could be heard.

15. Direct Speech : The traveller said, "Can you tell me the way to the nearest inn ?"

    "Yes" said the peasant ; "do you want one in which you can spend the night ?" "No", replied the traveller ; "I only want a meal."

    Indirect Speech : The traveller enquired of the peasant if he could tell him the way to the nearest inn. The peasant replied that he could, and asked whether the traveller wanted an inn in which he could spend the night. The traveller answered that he did not wish to stay there, but only wanted a meal.

EXERCISE

Change into indirect speech :

1. The old man said to his son, "What harm have I done to you ?"
2. She replied, "I'm going to walk where I like. We've got liberty now."
3. He said to the policeman, "Did you see a tall, fair man walking along the street yesterday ?"
4. The teacher said to the student, "Why did you misbehave in the class ?"
5. He said, "Mohan comes to college daily."

6. He said to us, "Are you coming to the meeting today ?"

7. Sita said, "What a pretty flower this is !"

8. He said, "How many brothers do you have?"

9. He said, "What a kind-hearted man you are !"

10. The old man said, "My sons, I am dying. I wish you get all my riches."

11. I said to him, "Let us go out for a walk."

12. The boy said to the Headmaster, "Good morning, Sir. Thank you for your kindness."

13. "Alas ! What shall I do ?" Cried Lucy, "I have lost my way in the snow."

14. The speaker said, "I entirely object to your proposal. Have you considered all that this proposal involves ? Gentlemen, I entreat you to be cautious."

15. She said to the King, "I have recognised my ring on the hand of the beggar who sits by the side of the garden. Father, send for him that we may find out how the ring came into his hands."

16. He said, "I have been waiting here for a long time."

17. Her mother said, 'You must go straight to your grandmother's cottage and do not loiter on the way. There is a wolf in the wood through which you are going ; but if you keep to the road he won't do you any harm. Now will you be a good girl and do as I tell you ?"

18. "You say", said the judge, "that the bag you lost contained one hundred and ten pounds." "Yes, my lord", replied the miser. "Then as this one contains one hundred pounds it cannot be yours."

19. The old man said to his friend, "Sit down and rest yourselves here on this bench. My good wife has gone to see what you can have for supper."

20. The teacher became angry with the student and said, "Why have you again disturbed the class in this way ? I have told you before that when I am speaking, you shall be silent. Leave the room and do not show me your face again today."

(C) Affirmative and Negative Sentences

The form of a sentence can be changed without changing the meaning. In this topic, we will study how affirmative sentences can be changed into negative sentences; and vice-versa.

Example 1 :   He was <u>doubtful</u> whether it was you.

                    He was <u>not sure</u> whether it was you.

1. **Mark out the word which can be changed into antonym.**

2. **Find out the exact opposite meaning of the word.**

3. **Rewrite the sentence using opposite (antonym) of that word.**

4. **Before using antonym use one negative word and complete the remaining part of the sentence.**

5. **Note that transforming the sentence from one form to another, the meaning of the sentence should not change.**

Conversion of (i) affirmative to negative and (ii) negative to affirmative.

Example 2 :   He was too weak to walk.

He was so weak that he could not walk.

(Here, in this example, the method of Remove 'too' is used to make sentence negative).

Example 3 :   As soon as he came, she went out.

No sooner did he come than she went out.

(Here, in this example, 'As soon as' is replaced by 'No sooner .... than' for converting the given sentence into negative).

Example 4 :   She is the tallest girl.

No other girl is as tall as she.

(Here, in this example, change of degree has been used to make sentence negative).

Make Negative :

1. It was doubtful whether it was you.

Ans. It was not sure that it was you.

2. I shall always remember your goodness.

Ans. I shall never forget your goodness.

3. I am a poor man.

Ans. I am not a rich man.

4. It is difficult to kill Medusa.

Ans. It is not easy to kill Medusa.

5. You are very brave.

Ans. You are not a coward.

6. Always speak the truth.

Ans. Never tell a lie.

7. Everybody will admit that English is a great language.

Ans. Nobody will deny that English is not a great language.

8. He is sometimes foolish.

Ans. He is not always wise.

9. She was too weak to run.

Ans. She was so weak that she could not run.

10. As soon as the teacher came, the boys stood up.

Ans. No sooner did the teacher come than the boys stood up.

11. He was the tallest boy.

Ans. No other boy was as tall as he.

12. He needs only love.

Ans. He needs nothing but love.

13. His mission was successful.

Ans. His mission did not fail.

14. It's a strange bird.

Ans. It is not a common bird.

15. It's empty.

Ans. There is nothing in it. It's not full.

16. I envy the mighty sleepers.

Ans. I do not like the mighty sleepers.

17. And the cranks are always with us.

Ans. And the cranks are never away from us.

18. Her husband was dead.

Ans. Her husband was not alive.

19. Both pictures would be correct.

Ans. Neither of the pictures would be incorrect.

20. You misunderstood me.

Ans. You did not understand me correctly.

21. He could only talk about his illness.

Ans. He could talk about nothing but his illness.

22. Be careful.

Ans. Don't be careless.

23. It's risky to navigate the ship in darkness.

Ans. It's not safe to navigate the ship in darkness.

24. The light was dim.

Ans. The light was not bright.

25. They differ in face and figure, food and clothing.

Ans. They are not similar in face and figure, food and clothing.

26. Its transit failed to produce the desired stamp.

Ans. Its transit did not produce the desired stamp.

27. The smoking habit was considered dangerous.

Ans. The smoking habit was not considered safe.

28. I'll remember that.

Ans. I won't forget that.

29. In this respect it closely resembles alcohol.

Ans. In this respect it does not differ from alcohol.

Make Affirmative

1. Sounds are not easy to translate into words.

Ans. Sounds are difficult to translate into words.

2. The suitcase wasn't heavy.

Ans. The suitcase was light.

3. These shelters were not permanent.

Ans. These shelters were temporary.

4. It can't be true.

Ans. It must be false.

5. He is not dull.

Ans. He is clever.

6. He cannot play.

Ans. He is unable to play.

7. She does not attend.

Ans. She fails to attend.

8. It is of no use.

Ans. It is useless.

9. There is nobody in the class.

Ans. There is hardly anybody in the class.

10. No one was present.

Ans. Hardly anyone was present.

11. He left no plan untried.

Ans. He tried every plan.

12. He is not happy.

Ans. He is unhappy.

13. He did not know it.

Ans. He knew little of it.

14. The light was not on.

Ans. The light was off.

15. She did not like it.

Ans. She disliked it.

16. There was nothing to see.

Ans. There was hardly anything to see.

17. I shall never break my promise.

Ans. I shall always keep my promise.

18. His sorrow knew no bounds.

Ans. His sorrow was boundless.

19. But Shankar paid no heed.

Ans. But Shankar was heedless.

20. This feeling did not prove to be far wrong.

Ans. This feeling proved to be nearly right.

21. He didn't know anything. OR

He knew nothing.

Ans. He was quite ignorant.

22. He will not be able to do it.

Ans. He will be unable to do it.

23. I can't wait any longer.

Ans. I can wait only for a while.

24. This was not the first struggle in the life of Shri. Bapat.

Ans. There had been several other struggles earlier in the life of Shri. Bapat.

25. I don't see anything wonderful about it.

Ans. I fail to see anything wonderful about it.

I see nothing wonderful about it.

26. We didn't have much money.

Ans. We had only little money.

27. But this does not complete the task of the village worker.

Ans. But this leaves the task of the village worker incomplete.

28. No two persons will think exactly alike.

Ans. Any two persons are bound to think somewhat differently.

29. The poet was not the only one who had heard the strange melody.

Ans. There were others besides the poet who had heard the strange melody.

30. This has never been more true than in our own age.

Ans. This is more true in our own age than it has ever been.

EXERCISE

Change the following Negative sentences into Affirmative sentences and vice-versa :

1. I shall not forget your kindness.
2. There is no smoke without fire.
3. He is sometimes foolish.
4. Calcutta is the biggest city in India.
5. The two friends are not unlike each other.
6. I am not puzzled by this.
7. He is not illiterate.
8. As soon as we entered the compartment, the train started.
9. I am not tired.
10. The knife is not sharp.
11. The brave alone deserves the fair.
12. Napolean was the greatest general of his time.
13. All men are mortal.
14. Only graduates need apply for the job.
15. This fact is too evident to require a proof.
16. These fishing nets are all the wealth I own.
17. No sooner did he see the tiger than he fled.
18. I am not so intelligent as you.
19. Old men are not always wise.
20. No one but a millionaire can afford such luxuries.

(D) Interchange of Interrogative and Assertive Sentences

A question is sometimes set, not for the sake of getting information, but to suggest the answer that the speaker or writer desires to give.

In such interrogatives, when the question is affirmative (see example No. 1), a negative answer is implied, and when the question is negative, (see example No. 2), an affirmative answer is implied.

Example 1 : Interrogative : A wounded spirit who can bear ?

Assertive : No one can bear a wounded spirit.

Example 2 : Interrogative : Is not blood thicker than water ?

Assertive : Blood is thicker than water.

Transformation of Assertive into Interrogative :

Example 3 : Assertive : We were not sent into the world to make money.

Interrogative : Were we sent into the world simply to make money ?

Auxi- Sub- Main

liary ject Verb

verb.

Following method may be adopted for changing the assertive sentence to interrogative :

1. **Mark out the auxiliary verb (or main verb)**

2. **Put the auxiliary verb (if there is no auxilliary verb but main verb) before the subject.**

3. **Place the subject after verb.**

4. **Use one negative word after subject, if there is no negative word in the first sentence. If there is a negative word in the first sentence, remove it.**

5. **Use main verb and remaining part of the sentence.**

6. **Use interrogative mark at the end of the sentence.**

Example 4 : Assertive : He <u>was</u> a villian to do such a deed.

Interrogative : Was he <u>not a villian</u> to do such a deed ?

(Negative word)

(Here negative word is used in an interrogative sentence as it is not there in the first sentence).

Example 5 : Assertive : Their glory can <u>never</u> fade.

Negative word

Interrogative : Can their glory <u>ever</u> fade ?

Here negative word is removed as it is there in the first sentence.

Make Interrogative :

1. He is a fool to do such a deed.

   Isn't he a fool to do such a deed ?

2. We should not waste money on useless things.

   Should we waste money on useless things ?

3. He is a very naughty boy.

   Isn't he a very naughty boy ?

4. No one can be so hard-hearted.
   Can anyone be so hard-hearted ?
5. We shall never forget our kind friend.
   Shall we ever forget our kind friend ?
6. We were not sent into the world simply to make money.
   Were we sent into the world simply to make money ?
7. I could not bear this injustice.
   Could I bear this injustice ?
8. There is nothing better than a busy life.
   Is there anything better than a busy life ?
9. Death does not make any difference between man and man.
   Does death make any difference between man and man ?
10. No one has seen the other face of the moon.
    Has anyone seen the other face of the moon ?
11. He is the person who got the first prize.
    Isn't he the person who got the first prize ?
12. I shall never forget those happy days ?
    Shall I ever forget those happy days ?
13. She is a good girl.
    Isn't she a good girl ?
14. There is nothing to forgive.
    What is there to forgive ?
    Is there anything to forgive ?
15. There is nothing nobler than love.
    Is there anything nobler than love ?
16. I can never forget your kindness.
    Can I ever forget your kindness ?
17. Everybody worships the rising sun.
    Who doesn't worship the rising sun ?
18. We should not waste our time in idle speculations.
    Should we waste our time in idle speculations ?

EXERCISE

Make Interrogative

1. I never asked you to do it.
2. I shall never forget those happy days.
3. The action of the sea is very powerful.
4. Coal mining has always been a dangerous occupation.
5. No one has found any kind of life on the moon.
6. Dr. Bhabha was a very extraordinary scientist.
7. He was a villain to do such a deed.

> 8. I can never forget your kindness.
> 9. The child is very lovely.
> 10. It is foolish of Sharma to throw up his job.
> 11. Their glory can never fade.
> 12. No one can build a house on sand.
> 13. We were not sent into the world simply to make money.
> 14. He was a villain to do such a deed.
> 15. There is no sense in doing that.
> 16. Health is more precious than wealth.
> 17. It is useless to preach religion to a hungry man.
> 18. The leopard cannot change its spots.
> 19. That is not the way a gentleman should behave.
> 20. You cannot please everybody.

Conversion of Simple Sentences Into Complex

A simple sentence can be converted into a complex sentence by expanding a word or phrase into a subordinate clause. The subordinate clause may be either (i) noun clause (ii) adjective clause or (ii) adverb clause depending upon the nature of the sentence to be converted.

Example 1 :   Simple   :   He hoped <u>to pass.</u> (Noun clause)

Complex   :   He hoped that he would pass.

While converting a simple sentence into a complex one, the following steps would be useful :

1. **Find out the word or phrase which can be expanded into a subordinate clause.**
2. **Rewrite the main sentence.**
3. **Use proper conjunction for making the sentence complex.**
4. **Use appropriate pronoun for subordinate clause for the second sentence.**
5. **Use proper verb in the second sentence, keeping in mind the tense of the first sentence.**
    In the above example, the main verb in the first part of the sentence is in the past tense, hence the verb in the second sentence is also kept in the past tense.
6. **Do not disturb the sequence of the tense.**
7. **Use the remaining part of the sentence as it is.**

Example 2 :   Simple   :   An honest man is the noblest work of God. (Adjective clause)

Complex   :   An honest man is the noblest work that God has created.

Follow the same method as mentioned in case of the first example and transform the sentence.

Example 3 :   Simple   :   He rises with his brother. (Adverb clause)

Complex   :   He rises when his brother rises.

Follow the same method as mentioned in case of the first example and transform the sentence into complex.

Make complex :

1. I was certain of passing the examination.

    I was certain that I would pass the examination.

2. I helped the poor man with a large family.

   I helped the poor man who had a large family.

3. Only on my friend's arrival we went to the circus.

   As soon as my friend arrived, we went to the circus.

4. He studied hard to pass the examination.

   He studied hard so that he might pass the examination.

5. Keep quiet, or you will be punished.

   Unless you keep quiet, you will be punished.

6. The uranium atom is hit by a neutron to split it.

   The uranium atom is hit by a neutron so that it may split.

7. The rock was too heavy for them to move.

   The rock was so heavy that they could not move it.

8. It was good for him to hear the siren.

   It was good that he heard the siren.

9. They did not know what to do .

   They did not know what they should do.

10. The engineers could not decide where to build the bridge.

    The engineers could not decide where they should build the bridge.

11. He made changes in the plan to make it more useful.

    He made changes in the plan so that it could be more useful.

12. He was very happy to get his appointment order.

    He was very happy when he got his appointment order.

13. The workers are agitating to get better wages.

    The workers are agitating so that they may get better wages.

14. Mercury is not easily visible but it may sometimes be seen without a telescope.

    Though Mercury is not easily visible, it may sometimes be seen without a telescope.

15. The train entered a tunnel but the lights did not go on.

    Though the train entered a tunnel, the lights did not go on.

16. We have not been told when to start the work.

    We have not been told when we should start the work.

17. This is the job for the artisans to complete today.

    This is the job which the artisans should complete today.

18. The Russians sent a rocket past Venus to know more about the planet.

    The Russians sent a rocket past Venus so that they could know more about the planet.

19. The directors were satisfied to see the scale model.

    The directors were satisfied when they saw the scale model.

20. The model car is put to severe tests to make it foolproof.

    The model car is put to severe tests so that it can be made foolproof.

21. He has gone to the U.S.A. to specialise in glass technology.

    He has gone to the U.S.A. so that he may specialise in glass technology.

22. Being tired, we returned home.

    We returned home because we were tired.

23. The boys stood up at the entry of the teacher.

    As soon as the teacher entered, the boys stood up.

24. I make a promise, only to keep it.

    If I make a promise, I keep it.

25. He is rich but he is not happy.

    Though he is rich, he is not happy.

26. Having finished his work, he left the office.

    When he finished his work, he left the office.

27. Here comes the girl with blue eyes.

    Here comes the girl whose eyes are blue.

28. Behave well or you will be punished.

    Unless you behave well, you will be punished.

29. The job was difficult but they did not give it up.

    Although the job was difficult, they did not give it up.

30. Any task well–accomplished gives satisfaction.

    Any task, which is well–accomplished, gives satisfaction.

Conversion of Complex Sentences into Simple

Complex sentences, containing noun clause, introduced by the conjunction "that", a Relative pronoun and Relative adverb, can be converted into simple ones by the substitution of nouns for such noun clauses.

Example 1 :   Complex   :   He confessed that he was guilty.

                   Simple   :   He confessed his guilt.

1. **Mark out the subordinate clause.**

2. **Remove the conjunction.**

3. **Remove the pronoun (It means remove the subject of the second sentence).**

4. **Use proper word which can be perfectly substituted.**

5. **Rewrite the remaining part of the sentence, if any.**

Example 2 :   Complex   :   Tell me when and where you were born.

                   Simple   :   Tell me the time and place of your birth.

(Follow the same method as mentioned in the first example)

Example 3 :   Complex   :   You are at liberty to criticise what I did.

                   Simple   :   You are at liberty to criticise my action.

(Follow the same method as mentioned in the first example).

Make Simple :

1. He finished his work and went home.

   After finishing his work he went home.

2. We must help those who are in need.

   We must help the needy.

3. He accepted that he was wise.

   He accepted his wisdom.

4. He was so poor that he could not pay his fees.

   He was too poor to pay his fees.

5. Though he is wealthy, he is not happy.

   In spite of his wealth, he is not happy.

6. He collects a lamp from the lamp room and goes to the pit head.

   Collecting a lamp from the lamp room, he goes to the pit head.

7. My friend arrived, and we went for a walk.

   On my friend's arrival, we went for a walk.

8. You must not be late or you will be punished.

   In the event of your being late you will be punished.

9. He worked hard but he did not succeed.

   In spite of his hard work, he did not succeed.

10. They were surprised when they saw the atomic reactor at Trombay.

    They were surprised to see the atomic reactor at Trombay.

11. The man ran so that he could catch the bus.

    The man ran to catch the bus.

12. We are proud that we have built the bridge in a year.

    We are proud to have built the bridge in a year.

13. They build fuel tanks so that they can store petrol for the trucks.

    They build fuel tanks to store petrol for the trucks.

14. The manager knows who he should order for the supply of the parts.

    The manager knows who to order for the supply of the parts.

15. The engineer advised the overseer what he should do in the situation.

    The engineer advised the overseer what to do in the situation.

16. The car is so expensive that I cannot buy it.

    The car is too expensive for me to buy.

17. Chlorine is added to water so that it may kill the germs.

    Chlorine is added to water to kill the germs.

18. The bank gave him a loan so that he may set up a factory.

    The bank gave him a loan to set up a factory.

19. A volcano gives plenty of warning before it is going to erupt.

    A volcano gives plenty of warning before erupting.

20. He must start early or he will not reach the office in time.

    Without starting early he will not reach the office in time.

21. She was unhappy because she had lost her passport.

    Having lost her passport, she was unhappy.

22. He expected that he would see his friend at the function.

    He expected to see his friend at the function.

23. As the invaders had gained the victory, they left at once.
     Having gained the victory, the invaders left at once.
24. Success often comes to people who work hard.
     Success often comes to hard – working people.
25. I know what he is thinking.
     I know his thoughts.
26. He tried hard but he could not mend the engine.
     In spite of trying hard he could not mend the engine.
27. The workers were so tired that they did not wake up.
     The workers were too tired to wake up.
28. What you really want is to sell the land.
     You really want to sell the land.
29. It has not only lengthened life but it has deepened its quality.
     Besides lengthening life, it has also deepened its quality.
30. Keep this in mind and act accordingly.
     Keeping this in mind, act accordingly.

EXERCISE

Make complex sentences :

1.   He was obviously a well–intentioned person.
2.   His father is not likely to punish him.
3.   They started with breaking a cup.
4.   Sudhir admitted his guilt.
5.   I have informed him of our success.
6.   The fog having lifted, the plane took off.
7.   Being an air hostess, she has seen most of the world.
8.   In spite of all precautions, the dam burst.
9.   Let us wait till his arrival.
10. He was too weak to walk.
11. On seeing the police, he ran away.
12. He bought his brother's property.
13. The reason for his arrest is still unknown.
14. Only on my friend's arrival we went for a walk.
15. At the end of the war, the United Nations was formed.
16. I saw a wounded bird.
17. This proves his innocence.
18. Hard–working pupils may win a prize.
19. Tell me the time and place of your birth.
20. The sun having set, we had better start for home.

Conversion of compound sentences into simple sentences :

The form of a sentence can be changed without changing the meaning. In this topic, we will study how compound sentences can be changed into simple sentences and vice–versa. A compound sentence can be transformed into a simple sentence by converting the coordinate clause into a phrase.

For example :

1. They worked hard, but they did not succeed. (Compound Sentence)

    In spite of hard work, they did not succeed. (Simple Sentence)

2. That dress can't be yours, for it is too big. (Compound)

    That dress is too big to be yours. (Simple)

3. You must stop consuming sugar or you cannot recover from this illness. (Compound)

    You must stop consuming sugar to recover from the illness. (Simple)

4. He completed his studies and went home. (Compound)

    Having completed his studies, he went home. (Simple)

5. He was tired, so he lay down. (Compound)

    Being tired, he lay down. (Simple)

EXERCISE

Rewrite the following compound sentences as simple ones :

1. They were ill, and, therefore, they could not attend the meeting.

2. He was careless, yet he passed.

3. Make haste or you will miss the train.

4. He is rich, but he is not satisfied.

5. They are poor, but they do not complain.

6. The manager took all precautions, yet the workers went on Strike.

7. The boiler suit cannot be yours, for it is too short.

8. He ate excessively, so he suffered from fatness.

9. Students must study hard or they will not pass.

10. Not only men, but women and children were also affected.

Complex Sentences into Compound sentences

A complex sentence can be transformed into compound sentence by changing the subordinate clause into a co-ordinate clause.

For example :

1. He takes regular exercise so that he may be fit. (Complex)

    He wants to be fit; therefore, he takes regular exercise. (Compound).

2. If you don't study sincerely, you will lose your position in the class. (Complex)

    Study sincerely or else you will lose your position in the class. (Compound)

3. Though I warned him again, the boy did not care to correct himself. (Complex)

    I warned the boy again, but he did not care to correct himself. (Compound)

4. Unless you improve your behaviour, you will be expelled from the institute. (Complex)

    Improve your behaviour or else you will be expelled from the institute. (Compound)

5. The mob dispersed as soon as the police arrived. (Complex)

    The police arrived and the mob dispersed. (Compound)

EXERCISE

Rewrite the following complex sentences as compound sentences :
1. In spite of the fact that we helped him in every way, he failed.
2. It is unfortunate that he has failed.
3. I like the book which my friend had presented to me.
4. We are quite sure that they will come here.
5. It began to rain heavily when I reached the institute.
6. He came first in his class is a matter of great pride.
7. I know that he is a liar.
8. I went home after I had finished my practicals.
9. He accepted the advice that I gave him.
10. He must suffer because he is mistaken.

Interchange of exclamatory sentences to Assertive Sentences

An exclamatory sentence can be transformed into an assertive sentence with the same meaning.

For example :
1. How attractive this picnic spot is !   (Exclamation)
   This picnic spot is very attractive.   (Assertion)
2. What a fool he is !   (Exclamation)
   He is a great fool.   (Assertion)

EXERCISE

Write the following exclamatory sentences as assertives carrying the same meaning :
1. How glad was I to be in my house once again !
2. What a beautiful scene this is !
3. How depressed he looks !
4. What a coward he is !
5. Oh if I were young again !

EXERCISE

Rewrite the following assertive sentences as exclamatory sentences :
1. They are very foolish.
2. It was a horrible accident.
3. It was avery tragic message.
4. She sings very sweetly.
5. He made a silly mistake.
6. His performance is very poor.

(E) Question Tag

The term Question Tag, which is also called 'tag question', is a phrase added at the end of a statement so as to turn it into a question in order to ascertain that the statement made is correct. To use a question tag is common practice in a conversation. This phrase is generally in the form of a short question. For example,

It is very cold, **isn't it ?**

or

He can't read, **can he ?**

The underlined phrases in the above examples are called as question tag or tag question.

The general pattern of question tag is –

(i)   Auxiliary verb + n't + subject, if the statement is positive and

(ii)   Auxiliary verb + subject, if the statement is negative.

It is a point to be noted that the subject of the question tag is always a pronoun but never a noun. It should also be noted that the question tag is negative when the main sentence is positive and it is positive when the main sentence is negative.

Some more examples of question tag are given below :

(1)   He is always ready, isn't he ?

(2)   She doesn't know him, does she ?

(3)   He is busy, isn't he ?

(4)   My mother sings well, doesn't she ?

(5)   It is pouring outside, isn't it ?

(6)   Ashok doesn't play well, does he ?

EXERCISE

Add question tag to the following :

(1)   He is wrong.

(2)   She doesn't argue.

(3)   They have purchased a plot.

(4)   His father is an engineer.

(5)   We are working hard.

(6)   She lost her temper.

(7)   My job is over.

(8)   They are shouting loudly.

(9)   The college is very big.

(10)   People are not lazy.

(F) Remove Too

The transformation of sentences concerns with the changing of sentence from one grammatical form to another without altering its sense or meaning.

Removing an adverb 'too' is one of the methods of transforming a sentence. There are different ways and means of transforming different types of sentences by using the adverb 'too.'

(i)   So .... that .... not/cannot ....

Example : You are too young to understand this.

After removing too the sentence becomes :

You are **so** young **that** you **cannot** understand this.

In this type, the main verb from the original sentence takes the form in a negative sense after removing the adverb 'too'.

(ii) More than necessary.

  Example : Too many cooks spoil the broth.

  After removing too the sentence becomes : More than necessary cooks spoil the broth.

(iii) Excessively.

  Example : Do not be too eager to praise.

  After removing too the sentence becomes : Do not be excessively eager to praise.

In order to attempt this question the following steps be followed :

1. **First of all mark out the main sentence and phrase.**

2. **Use 'so' in the place of 'too' and rewrite the sentence.**

3. **Use one suitable conjunction 'that or which' – whichever is applicable.**

4. **Use one suitable pronoun, according to the sense of the first sentence to work as the subject of the second sentence.**

5. **Main sentence**   **Use in Second Sentence**

  (a) Present Tense   **Cannot**

  (b) Past Tense.   **Could not.**

6. **Use the remaining part of the sentence.**

7. **In some sentences, the second sentence can be changed to Passive Voice, to make the meaning clear.**

For removing the adverb 'too', [questions can be asked in different ways]. The following are the methods by which the question can be asked :

(A) Remove 'too'

(B) Use 'so'

(C) Make complex

(D) Make negative.

It means that, in removing the adverb 'too' we are expected to follow all the above conditions. Only then, the given question can be transformed without changing its sense.

Remove 'too'

 1. This tree is too high for me to climb.

Ans. This tree is so high that I cannot climb it.

 2. The motor–cyclist was going too fast for the police–car to overtake it.

Ans. The motor–cyclist was going so fast that he could not be overtaken by the police-car the police-car could not overtake him.

 3. He was sobbing too deeply to answer.

Ans. He was sobbing so deeply that he could not answer.

 4. This news is too good to be true.

Ans. This news is so good that it cannot be true.

 5. He is too old to walk.

Ans. He is so old that he cannot walk.

 6. He is too proud to beg.

Ans. He is so proud that he will not beg.

7. These mangoes are too cheap to be good.

Ans. These mangoes are so cheap that they cannot be good.

8. She was too weak to move about in the house.

Ans. She was so weak that she could not move about in the house.

9. He was too tired to work.

Ans. He was so tired that he could not work.

10. He was too busy to pay attention to his son's education.

Ans. He was so busy that he could not pay attention to his son's education.

11. It is too late for us to do anything this evening.

Ans. It is so late, that we cannot do anything this evening.

12. The light was too dim for him to take aim.

Ans. The light was so dim that he could not take aim.

13. The slippers were too large for her to wear properly.

Ans. The slippers were so large that she could not wear them properly.

14. The ice was too thick to break into pieces.

Ans. The ice was so thick that it could not be broken into pieces.

15. The class is too crowded to be controlled.

Ans. The class is so crowded that it cannot be controlled.

16. Akbar was too shrewd to be deceived by the words of the courtiers.

Ans. Akbar was so shrewd that he could not be deceived by the words of the courtiers.

17. The summer is too severe to change the place of the regiment.

Ans. The summer is so severe that the place of the regiment cannot be changed.

18. He was too excited to eat.

Ans. He was so excited that he could not eat.

19. It was too late for them to save their fishing gear.

Ans. It was so late that they could not save their fishing gear.

20. The stone was too heavy to move.

Ans. The stone was so heavy that it could not be moved.

21. She is too fat to run.

Ans. She is so fat that she cannot run.

22. You love your girls too much to neglect such an acquaintance.

Ans. You love your girls so much that you cannot neglect such an acquaintance.

23. This fact is too evident to require proof.

Ans. This fact is so evident that it does not require proof.

24. The example is too difficult for him to solve.

Ans. The example is so difficult that he cannot solve it.
The example is excessively difficult for him to solve.

25. It was too cold for me to go out.

Ans. It was so cold that I could not go out.

26. It is too dark to see.

Ans. It is so dark that nothing can be seen.

27. They were too late to avoid another disaster.

Ans. They were so late that they could not avoid another disaster.

28. It is never too late to mend.

Ans. It is never so late that it cannot be mended.

29. He is too poor to afford a car.
Ans. He is so poor that he cannot afford a car.
30. It was too expensive for her to buy.
Ans. It was so expensive that she could not buy it.

EXERCISE

Remove 'too' from the following sentences :

1. She was too old to walk fast.
2. Her sorrow was too deep to be wept away.
3. He is too honest to deceive his neighbours.
4. The news is too absurd to believe.
5. The resistance of air was too great for the aeroplane to fly at a high speed.
6. The question is too difficult to solve.
7. They are too poor to pay for shelter.
8. The crowd was too fierce to control.
9. Sheela was too proud to talk to us.
10. Shekhar is too fat to run.
11. It is too far to go on foot.
12. The boy was too clever to be taught.
13. He is too old to work.
14. The teacher speaks too fast to be understood.
15. The book is too lengthy to complete within a day.
16. The news is too bad to be false.
17. He is too clever to be deceived.
18. It is too hot to drink.
19. He is too poor to buy books.
20. The weather is too bad for us to go out.
21. The building is too old to be repaired.
22. We reached the station too late to catch the train.
23. He is too ill to go to school today.
24. He is too good to do harm to anybody.
25. He was too reserved to make any friends.

Change of Degree :

Comparison of Adjective

Change of degree has a basic concern with the part of speech viz. adjective. Hence the concept of what is an adjective must be very clear in the minds of the students. An Adjective can be defined as **"A word used with a noun to describe, or point out, the person, animal, place or thing which the noun names, or to tell the number or quantity."** In other words, it can be stated as a word used with a noun to add something to its meaning.

For example : He is a **happy** boy.

Here, the word 'happy' shows what kind of boy he is. However, the extent of happiness of the boy is not clear.

He is happier than Hari.

In this sentence, however, the adjective 'happier' tells us that his happiness is greater than Hari's happiness, when compared with one another.

He is the happiest boy in the class.

Here, the adjective 'happiest' tells us the highest degree of happiness.

It can be seen from the above examples that the adjective 'happy' changes its form to happier or, happiest to show the comparison. They are called Degrees of Comparison.

1.　The adjective 'happy' is said to be in the Positive Degree.

2.　The adjective 'happier' is said to be in the Comparative Degree.

3.　The adjective 'happiest' is said to be in the Superlative Degree.

The Positive Degree of an adjective is in its simple form. It denotes a mere existence of some quality of what we speak. It is used when no comparison is made.

The Comparative degree of an adjective denotes a higher degree of the quality than that of the positive, and it is used when two things (or sets of things) are compared.

The Superlative Degree of an adjective denotes the highest degree of the quality, and is used when more than two things are compared.

Formation of Comparative and Superlative forms of adjectives :

1.　Most of the adjectives of one syllable and some of more than one syllable form the comparatives by adding 'er' and the Superlative by adding 'est' to their Positive form.

Example :

Positive	Comparative	Superlative
Tall (one syllable)	Taller	Tallest
Happy (More than one syllable).	Happier	Happiest

2.　Adjectives of more than two syllables, and also many of those with two syllables, form their Comparatives by using the adverb *more* and their Superlatives by using the adverb *most* with the positive form.

Example :

Positive	Comparative	Superlative
Beautiful	more beautiful	Most beautiful
Useful	more useful	Most useful

Irregular Comparison

There are also certain adjectives which are termed as irregular; that is, their comparative and superlative are not obtained from the Positive. In other words, the comparatives and superlatives are altogether different from their positives.

Example :

Positive	Comparative	Superlative
Good, well	better	best
Bad, evil, ill	worse	worst
Much	more	most
Many	more	most
Far	farther	farthest
Little	less, lesser	least
Fore	former, further	foremost, first, furtherest

The following method be adopted for transforming sentences from one degree to another degree.

1.　**Mark out the adjective from the given sentence.**

2.　**Find out the degree of the adjective, and write it.**

3.　**Interchange the places of both the nouns.**

4. Use an appropriate verb without changing the tense of the second sentence.
5. Use one negative word before the adjective, if it is not used in the first sentence. If there is a negative word in the first sentence, remove that negative word.
6. Use the adjective in the required degree.
7. After using the adjective, use the conjunction 'than.'
8. Keep the remaining part of the sentence as it is.

Example :

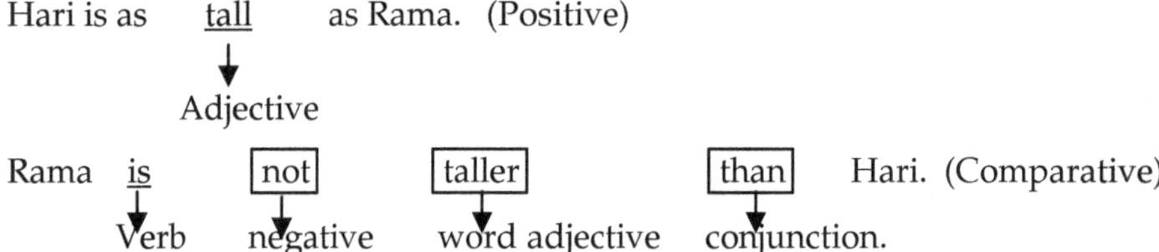

Interchange of the Degrees of Comparison

The following example shows that it is possible to change the degree of comparison of an adjective into other degrees like positive degree and superlative degree, without changing the meaning of the sentence. This can be well illustrated by taking an example of a sentence in the comparison.

Change from Positive to Comparative :

The following method may be adopted for transferring sentences from positive to comparative.

1. Find the adjective from the given sentence.
2. The last noun mentioned in the sentence is to be taken in the beginning as the subject for the Comparative Degree.
3. Use the proper verb, without changing the tense.
4. Use the adjective, in the required degree, i.e., in comparative.
5. After using the adjective, it is necessary to use the conjunction 'than'.
6. In case the positive degree starts with 'No other', it is necessary to use 'Any other' after the conjunction in the Comparative Degree.
7. Use the noun which comes in the beginning of the positive sentence.
8. Use the remaining part of the sentence, as it is.

Example :

Positive   No other city in India is as beautiful as Mumbai.

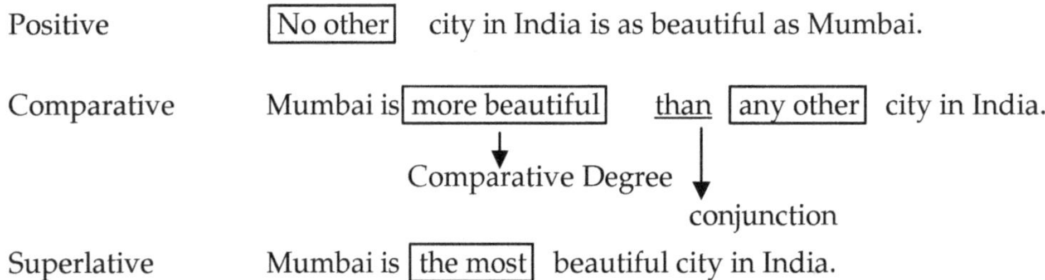

Superlative   Mumbai is the most beautiful city in India.

In the example mentioned above, comparison is done between one city and the remaining cities in India.

Change from Comparative to Superlative :

The following methods may be adopted for transferring sentences from comparative to superlative.

A. 1. Use the same subject, as it is used in the Comparative Degree.
  2. Use the proper verb without changing the tense.

3. Use 'the'.
4. Use the adjective in the Superlative Degree.
5. Use the noun.
6. Use the remaining part of the sentence.

Example :

In this example, comparison is done between a few and the remaining cities in India.

Positive | Very few | cities in India are as beautiful as Mumbai.

Comparative   Mumbai is more beautiful than | most other | cities in India

Superlative   Mumbai is | one of the | most beautiful cities in India.

B. 1. Find the adjective as well as its degree from the given sentence.
    2. The last noun mentioned in the sentence is to be taken in the beginning as the subject for the comparative degree.
    3. Use the proper verb, without changing the tense.
    4. Use the adjective in the required degree.
    5. After using the adjective, it is necessary to use the conjunction 'than'.
    6. In case the positive degree starts with "very few", it is necessary to use "most other" after the conjunction in the comparative degree.
    7. Use the noun which comes in the beginning of the positive sentence.
    8. Use the remaining part of the sentence.

Change of Degree from comparison to positive (when the comparison is between two objects).

    1. Ram is cleverer than Hari.
Ans. Hari is not as clever as Ram.
    2. Seeta is taller than Geeta.
Ans. Geeta is not as tall as Seeta.
    3. Plastic is cheaper than rubber.
Ans. Rubber is not as cheap as plastic.
    4. It's much more interesting than your album of film stars.
Ans. Your album of film stars is not as interesting as this.
    5. Soft-footed animals are a little more difficult to track than hard-footed ones.
Ans. Hard–footed animals are not as difficult to track as soft-footed ones.
    6. Nephews are worse than nieces.
Ans. Nieces are not as bad as nephews.
    7. Light music is more popular than classical music.
Ans. Classical music is not as popular as light music.
    8. Iron is more useful than copper.
Ans. Copper is not as useful as iron.
    9. Astrologers are more popular than astronomers.
Ans. Astronomers are not as popular as astrologers.
    10. I could see better than I had ever seen in my life.
Ans. I had never in my life seen as well as I could see then.

Change of Degree from superlative to comparative and positive.

When the comparison is between (a) One and remaining (b) Few and remaining.

    (a) One and remaining
    1. Ram is the cleverest boy in the class. (Superlative)
        Ram is cleverer than any other boy in the class. (Comparative)
        No other boy in the class is as clever as Ram. (Positive)

2. Calcutta is the largest city in India. (Superlative)
   Calcutta is larger than any other city in India. (Comparative)
   No other city in India is as large as Kolkata. (Positive)
3. Alfred was the best assistant in the shop. (Superlative)
   Alfred was better than any other assistant in the shop. (Comparative)
   No other assistant in the shop was as good as Alfred. (Positive)
4. Television is the quickest means of reaching out to our people. (Superlative)
   Television is quicker than any other means of reaching out to our people (Comparative)
   No other means of reaching out to our people is as quick as television. (Positive).
5. Onam is the biggest festival in Kerala. (Superlative)
   Onam is bigger than any other festival in Kerala. (Comparative)
   No other festival in Kerala is as big as Onam. (Positive)
6. Spring is the lovelist season. (Superlative)
   Spring is lovelier than any other season. (Comparative)
   No other season is as lovely as spring. (Positive)
7. Helen was the most beautiful woman. (Superlative)
   Helen was more beautiful than any other woman. (Comparative)
   No other woman was as beautiful as Helen. (Positive)
8. Iron is the most useful metal. (Superlative)
   Iron is more useful than any other metal. (Comparative)
   No other metal is as useful as iron. (Positive)
9. Village work is the best form of recreation. (Superlative)
   Village work is better than any other form of recreation. (Comparative)
   No other form of recreation is as good as village work. (Positive)
10. The most important of all this services was his contribution in the field of education. (Superlative)
    His contribution in the field of education was more important than any other of his services. (Comparative)
    No other of his services was as important as his contribution in the field of education. (Positive)

(b) Few and remaining :
1. Ram is one of the cleverest boys in the class. (Superlative)
   Ram is cleverer than most other boys in the class. (Comparative)
   Very few boys in the class are as clever as Ram. (Positive)
2. Swimming is one of the healthiest exercises. (Superlative)
   Swimming is healthier than most other exercises. (Comparative)
   Very few other exercises are as healthy as swimming. (Positive)
3. Mumbai is one of the biggest cities in India. (Superlative)
   Mumbai is bigger than most other cities in India. (Comparative)
   Very few cities in India are as big as Mumbai. (Positive)
4. Einstein is one of the greatest scientists of modern times. (Superlative)
   Einstein is greater than most other scientists of modern times. (Comparative)
   Very few other scientists of modern times are as great as Einstein. (Positive)
5. Birbal was one of the most famous ministers of Emperor Akbar. (Superlative)
   Birbal was more famous than most other ministers of Emperor Akbar. (Comparative)
   Very few other ministers of Emperor Akbar were as famous as Birbal. (Positive)
6. She is one of my best friends. (Superlative)
   She is better than most of my other friends. (Comparative)
   Very few of my friends are as good as she is. (Positive)

7. Mr. Bennet was among the earliest of those who waited on Mr. Bingley. (Superlative)
    Mr. Bennet was earlier than most others who waited on Mr. Bingley. (Comparative)
    Very few others who waited on Mr. Bingley were as early as Mr. Bennet. (Positive)
8. Curiosity is one of the noblest instincts of man. (Superlative)
    Curiosity is nobler than most other instincts of man. (Comparative)
    Very few other instincts of man are as noble as curiosity. (Positive)
9. He was one of the happiest farmers in his town. (Superlative)
    He was happier than most other farmers in his town. (Comparative)
    Very few other farmers in his town were as happy as he. (Positive)
10. This is one of the best books on the subject. (Superlative)
    This is better than most other books on the subject. (Comparative)
    Very few other books on the subject are as good as this. (Positive)

EXERCISE

Change the degree of the following sentences :

1. Madras is one of the biggest cities in India.
2. Sir John Cochrane was the most formidable rebel.
3. Gold is the most precious of all metals.
4. No leader in India was as popular as Mahatma Gandhi.
5. India is one of the hottest countries in the world.
6. No other metal is as useful as iron.
7. In the Grammar School she was the brightest of all the girls.
8. Riding is the best kind of exercise.
9. Einstein is one of the greatest scientists of the modern times.
10. Mumbai is one of the richest towns in India.
11. Everest is higher than all other peaks.
12. His father is richer than all other men in the town.
13. The aeroplane flies faster than do the birds.
14. The rose is the sweetest of all flowers.
15. The Kohinoor is one of the largest of diamonds.
16. The air of plains is not so cool as that of hills.
17. A wise enemy is better than a foolish friend.
18. Milk is the most nourishing food.
19. This is the most interesting novel I have read.
20. The Ganges is the longest river in India.

EXERCISE

Use the following adjectives in sentences of your own.

accountability	efficient	luxurious	rigorous
aeronautical	embracing	Measurable	Tiniest
Appreciative	enthusiastic	Mechanism	Traditional
Arrant	Ghastly	Multinational	tragic
atmospheric	Glamorous	Organic	Transparent
breezier	Hostile	Photographic	Unanimous
Despondent	Humanitarian	Provincial	unintelligivly
destructive	incomparable	racist	vulgar
disable	Inevitable	Rhythmic	wispy

(G) Use of Articles

In modern grammar, the term determiner is frequently used. A determiner is a word that comes before a noun to limit its meaning. The determiners are broadly classified into :

(i)   Articles,
(ii) Demonstrative adjectives,
(iii) Possessive adjectives and,
(iv) Adjectives of quality.

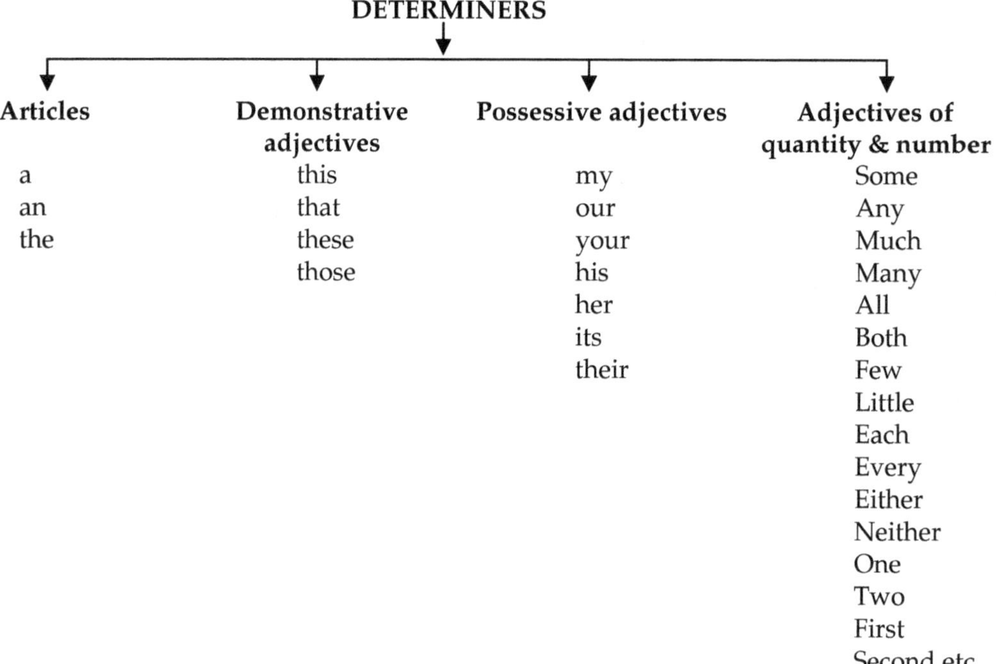

DETERMINERS

Articles	Demonstrative adjectives	Possessive adjectives	Adjectives of quantity & number
a	this	my	Some
an	that	our	Any
the	these	your	Much
	those	his	Many
		her	All
		its	Both
		their	Few
			Little
			Each
			Every
			Either
			Neither
			One
			Two
			First
			Second etc.

(i) Articles : Articles are of two types :

Articles

Indefinite   Definite

The articles 'a' and 'an' are called indefinite articles. They are called indefinite because they do not specify a person or a thing spoken of.

For example :

(i)   A teacher guides many students.

(ii) Take an umbrella for protection from rain.

In the above sentences, the words teacher and umbrella are nouns preceded by the articles 'a' and 'an' respectively. They do not specify any particular teacher or umbrella but any teacher or any umbrella. The article 'the' is called as a definite article as it is used to denote something specific in the context.

1. Indefinite Article

(a) The choice between **a** and **an** is determined by the sound with which the noun begins. Before a singular countable noun beginning with a consonant-sound, the article 'a' is used while before the singular countable noun beginning with a vowel - sound, the article 'an' is used.

For example :

Singular nouns beginning with a consonant sound	Singular nouns beginning with a vowel sound
A book	An engine
A pen	An enemy
A table	An hour
A chair	An honest man
A teacher	An elephant
A pencil	An instrument
A duster	An umbrella

The nouns under Col. 1 are preceded by the indefinite article 'a' while those under Col. 2 are preceded by the indefinite article 'an'.

The words like hour, honest though begin with a consonant, the initial consonant 'h' is not pronounced i.e. it is silent. As a result, their pronunciation begins with a vowel sound; hence they are preceded by the article 'an'. The words like European, university although begin with a vowel, the initial vowel is not pronounced like the word, 'you'. Similarly, in the word like 'one', the initial vowel 'w' and hence takes the article 'a'.

(b) The articles a/an are used to denote one.

    For example :   We can visit Pune, if we get **a** days' holiday.

                        Twelve inches make **a** foot.

                        Not **a** word was said by him.

                        A word to the wise is enough.

(c) Articles a/an are used in the vague sense as ______

    For example : One morning, **a** student visited my house. Here 'a student' means not a particular student.

(d) They can also be used in the sense of 'any' i.e. to single out **an** individual as the representative of a class.

    For example : A student should respect his teacher.

(e) Indefinite articles 'a' and 'an' are used to convert a proper noun into a common noun.

    For example :   A Daniel comes to judgement.

                        The word 'A Daniel' stands for a very wise man.

2. Definite Article :

'The' is called as a Definite article because it points out some particular person or thing in the context :

Uses of the Definite Article 'the'.

The Definite Article 'the'

(1) When we speak of a particular person or thing or the one already referred to.

    For example :   We like **the** fellow.

                        The book we want is not available.

                        Let me go to **the** club.

In the above sentences, the words 'fellow', 'book' and 'club' are particular in the context.

(2)　When a singular noun is meant to represent a whole class.

For example :　**The** horse is a useful animal.

The word 'horse' in the above sentence represent the whole class of horses.

The two nouns viz. 'man' and 'woman' used in the general sense to denote the whole class do not take any article.

For example :　Man is an intelligent being.

(3)　With the names of gulfs, rivers, seas, oceans, groups of islands and mountain ranges.

For example :　The Indian Ocean, the Himalayas, the Laccadive Islands, the Ganges.

(4)　Before the names of certain books.

For example :　The Ramayan, the Vedas.

(5)　 Before nouns which are the names of the things unique of their kind.

For example :　The Sun, the sky, the moon, and the earth.

(6)　'The' is used before a proper noun only when it is qualified by an adjective or a defining adjectival clause.

For example :　The Mr. Patil whom you met yesterday is my brother.

(7)　With superlatives.

For example :　This is **the best** book I have ever read.

(8)　With ordinals.

For example : She was **the first** woman to arrive.

(9)　Before musical instruments.

For example : She can play **the** flute.

(10)　Before an adjective when the noun is understood.

For example : The rich are never with us.

(11)　Before a noun (with emphasis) to give the force of superlative.

For example : The verb is **the word** (the main word) in a sentence.

(12)　As an adverb with comparatives.

For example : The more we get, **the** more we want.

Omission of Articles

No article is used –

(1)　No article is used before a noun when used in its widest sense.

For example :　Man is mortal.

(2)　before material nouns.

For example :　**Silver** is a precious metal.

　　　　　　　　Tea is grown in India.

(3)　before proper nouns.

For example :　**Kolkatta** is a big city.

　　　　　　　　Mumbai is the Capital of Maharashtra.

　　　　　　　　Newton was a great scientist.

Article 'the' is, however, used before a proper noun when it is used in the sense of common nouns.

For example : Mumbai **is the Manchester** of our country.

(4)　before Abstract nouns used in the general sense.

For example : Honesty is the best policy.

The article is however, used before an Abstract noun, when it is qualified by an adjective or an adjectival phrase or clause.

For example : We cannot forget **the kindness** with which they treated us.

(5) before languages.

> **For example :**   He is studying English.
> We speak Marathi.

(6) before the name of a college, school, temple, bed, table, hospital, market, prison, when these places are visited or used for their primary purpose.

> **For example :** We learnt Marathi at **School**
> We visit **temple** daily
> They stay in **bed** till early morning.

The article 'the', is however, used with these words when we refer them as a definite place, building or object rather than a place, building or object where the normal activity goes on.

> **For example :** **The** school is near my house.
> I met him at **the** temple.
> **The** bed is neat and clean.

(7) before names of relations, like mother, father uncle, aunt and also cook and nurse meaning 'our cook' or 'our nurse'.

> **For example :** **Father** has come.
> **Mother** wants to see him.
> **Cook** has left home.

(8) before predicative norms denoting a unique position i.e. a position that is normally held by one person only at one time.

> **For example :** He is elected **president** of the Council.
> Mr. Patil became **principal** of the College in 1994.

(9) certain phrases consisting of transitive verb followed by its object.

> **For example :** To catch fire, to give battle, to send word, to lose heart, to leave home etc.

(10) The article is omitted in certain phrases consisting of a preposition followed by its object.

> **For example :** At home, in hand, at daybreak, at noon, at sunset etc.

EXERCISE 1

Use 'a' or 'an' wherever necessary and rewrite the following sentences :

1. _______ spade is _______ instrument.
2. _______ potato is _______ vegetable.
3. _______ cigar is made of _______ tobacco.
4. _______ oil refinery is _______ place where _______ crude oil is purified.
5. We can write _______ letter in _______ ink.
6. _______ magnet attracts _______ piece of _______ iron.
7. What is _______ volcano ? It is not _______ mountain of _______ fire as most of us think.
8. _______ coat is made of _______ wool.
9. _______ cloud of _______ smoke hangs over the crater, out of which _______ molten rock pours.
10. _______ butter and _______ cheese are made from _______ milk.
11. _______ water soaks slowly downward through _______ sand and _______ gravel.
12. In _______ ordinary electric power station we burn _______ fuel to get _______ heat.
13. Blard's first apparatus was made of _______ old tea–chest _______ biscuit tin _______ thick lens and _______ toy electric motor.

14. Galileo was _______ lecturer in _______ mathematics at _______ university in _______ town called Pisa in _______ Italy.
15. A dog is _______ animal.
16. _______ chair is made of wood.
17. _______ eye is blue or brown.
18. _______ fish can swim.
19. _______ cow gives milk.
20. _______ picture is pretty.
21. _______ apple grows on _______ tree.
22. We can make _______ cake with flour, milk and _______ egg.
23. A writer writes _______ book.
24. Copper is _______ useful metal.
25. He returned after _______ hour.
26. I first met him _______ year ago.
27. Yesterday _______ European met me.
28. He came without _______ umbrella.
29. She is _______ untidy girl.
30. I bought _______ horse, _______ ox and _______ buffalo.

EXERCISE 2

Rewrite the following sentences inserting articles (a, an, or the) wherever necessary :

1. Only best quality is sold by us.
2. What kind of bird is that ?
3. It was proudest moment of his life.
4. April is fourth month of year
5. Man cannot live by bread alone.
6. Umbrella is of no use against storm.
7. Iron is metal.
8. There is fly in Lemonade.
9. Some birds can fly very high in sky.
10. I should like a house in country.
11. He went to fetch pale of water.
12. Do you prefer book of Engineering science ?
13. Youngest brother is at school now. If you go to school by bus, you will be just in time to see him.
14. Book on that shelf is interesting one.
15. He took up swimming as sport.
16. Telex looks like large typewriter.
17. Yesterday he met man, man was engineer.
18. Specification is a secret document passed on to stylist and engineers stylists study document very carefully.
19. Sun has risen but moon can still be seen in sky
20. My friend took me to site of dam.
21. I saw mechanical shovel at work site.
22. He has become politician.

23. Concrete that is used for building this road is not good.
24. Water that we use is free from germs.
25. Vapour when cooled turns into liquid.
26. Mason fell off scaffolding. Immediately mason was rushed to hospital.
27. Boy took pen from table and wrote letter.
28. River was spanned by iron bridge.
29. Moon did not rise till after ten.
30. There is nothing like staying at home for comfort.

(ii) Demonstrative Adjectives

Demonstrative adjectives show a person or thing spoken about. They answer the question 'which' ?

The demonstrative adjectives **this** and **that** are used with singular nouns, while, 'these' and 'those' are used with plural nouns.

For example :   **This** student is stronger than Rama.

That lady is intelligent.

These fruits are sweet.

Those persons must be rewarded.

(iii) Possessive Adjectives

Following are the possessive adjectives which belong to the group of determiners :

(My, our, your, his, her, its, their)

For example :   This is **my** pen.

(Possessive adjectives are formed from pronominal adjectives)

The word 'his' can be used both as an adjective as well as a pronoun.

For example :   This is his pen. (Possessive Adjective)

This pen is his. (Possessive pronoun)

(iv) Adjectives of quantity and number

Following determiners belong to the group of adjectives of quantity and number.

Some, any , much, many, all, both, few, little, each, every, either, neither, one, two, first, second etc.

Some : 'Some' is used with countables as well as uncountables in affirmative sentences. When it is used with a countable noun, it is the plural of 'an' and means 'an unknown number' of the noun. When it is used with an uncountable noun, it means 'an unknown quantity' of that noun.

For example : Some (of these) chemicals are used for protecting plants. (unknown number)

Some (of the) oil is used for fuel. (unknown quantity).

Any : Any is used with countables and uncountables in interrogative and negative sentences in the sense of one or some.

For example :   Is there any student here ? (countables)

There are hardly any students in the class (countables)

Are there any hills in that part ? (uncountable)

There is hardly any material left of the work (uncountable)

EXERCISES 3

Use 'some' and 'any' where required in the following sentences and rewrite them :
1. There is no student in the class.
2. Is there petrol left in the tank ?
3. Are there such planes that travel as fast as sound ?
4. The supervisor found tracks near the dam site.
5. Put more fruits in the bag.
6. "Is there progress in the construction work ?" "Yes, there is ; but it is not up to the mark."
7. Do not make noise. Let me have sleep. I did not have last night.
8. They have not got time to do more work.
9. Some scraping will have to be done to level the ground.
10. Do you expect to get money from the office ? If yes, I would like to borrow from you.

Much : Much is used with **uncountable** nouns in interrogative and negative sentences. It is not used by itself in affirmative sentences. Instead, **plenty of, a lot of, a great deal of** are used.

Example :   How **much** sugar is produced in India ? (uncountable)

Not **much** oil is produced in our country. (uncountable)

A lot of oil is produced in Middle East.

This is too much for him; he cannot consume all of it.

Many : Many is used with countable nouns in interrogative and negative sentences.
It can be used by itself in affirmative sentences. However, the use of plenty of, a lot of, instead of many is preferred.

Example :   How **many** people eat this kind of food ?

Not many eat this kind of good.

Plenty of people eat this kind of food.

A good many people have not liked to change.

Little, a little, the little :
(A) The adjective **little** has a negative meaning when it is used without article (a, the). Here, little gives the meaning as 'hardly any' or 'not much'.

For example :   1.   He has **little** influence over his friends.

2.   He has **little** appreciation of good work.

(B) **A little** has a positive meaning when it is used with article 'a'. Here 'a little' gives the meaning as 'Some though not much.'

For example :   A little tact would have saved him.

A little knowledge is a dangerous thing.

(C) **The little** has meaning of not much but there is some.

For example :   The little knowledge of engineering that he possessed helped him in his profession.

(Here the meaning is that the knowledge of engineering he possessed was not much ; but whatever knowledge he possessed helped him a good deal)

Few, a few, the few :
1. Few means not many or hardly any and has a negative meaning.
2. A few mean's some and it has a positive meaning and is opposed to 'none.'
3. The few means not many; but there is some.

EXERCISE 4

Insert much, many, little, a little, the little, few, a few, the few, in the following sentences wherever required and rewrite them :

1. Not people have telephones in India.
2. Only copies of the book are left with the book seller : the remaining have been sold out .
3. Grain they had was damaged by rains.
4. Precaution is necessary in handling that instrument.
5. Persons can keep a secret.
6. Men are free from faults.
7. Care should have prevented accident.
8. People are so hopeless as drunkards.
9. Clothes we had were all torn.
10. Public libraries are not well equipped.
11. Knowledge of English that he has is likely to be very useful to him in Engineering field.
12. It is a question of spending rupees.
13. When I met him years after he looked old.
14. I am not interested in coal minimising. I know about it.
15. How work can be finished today ?
16. The job is very dangerous workmen would be willing to do it up.

Insert in the blanks suitable adjectives of quantity and number and rewrite them :

17. ________ hints on essay-writing are quite to the point.
18. ________ months that he spent in Mahableshwar did him a lot of good.
19. In ________ words I expressed my gratitude.
20. ________ influence that he has, he uses to get employment.
21. ________ knowledge of carpentry he possessed helped him to get the job.
22. He showed ________ concern for his brother.
23. He showed ________ mercy to a poor person.
24. How ________ engineers are useful for the supervision.
25. Atoms are very small ________ too to be seen.
26. It seems there is very ________ oxygen on other plants.
27. After ________ minutes the mist cleared.
28. ________ days, rest is needed.
29. ________ days that are left to him he spends in solitude.
30. ________ men reach the age of one hundred years.

Modifiers

A word or phrase that modifies another word or phrase is called as a modifier. Most of the adverbs are formed from adjectives by adding suffix 'ly'.

For example :   **Adjective**   **Adverb**

      Frank   Frankly

      Happy   Happily

Adverbs have two important functions to perform. They are :

1. Adverbial in a sentence and
2. Modifier of adjectives, adverbs and other phrases.

(H) Prepositions

A preposition is a word placed before a noun or a pronoun to show in what relation the person or thing denoted by it stands in regard to something else. In other words, it can be stated as "that which is placed before".

Kinds of Prepositions

Following are the kinds of prepositions :

1. Simple Prepositions   (by, for, from, in, on, out, etc.).
2. Compound Prepositions   (About, along, around, behind etc.).
3. Phrase Prepositions   (in addition to, in case of, in front of etc.).
4. Participle Prepositions   (concerning, during, regarding etc.).

1. Prepositions usually precede the words they govern or control.

   He told us **about** the course.

   The boy **in** the corner is my friend.

2. Prepositions indicate time.

   He worked **in** the factory **from** 1992 **until** 1995.

3. Prepositions indicate positions.

   He noticed the aeroplane **above** that building.

4. Prepositions indicate direction.

   They left **for** Pune this morning.

5. Prepositions indicate the direction from the point of starting.

   The road runs **from** Pune to Mumbai.

6. Certain prepositions indicate direction.

   The road runs **across** the forest.

Relations expressed by Prepositions

Following are some of the most common relations indicated by prepositions.

1. **Place** (about, across, against, among etc.).

   He ran **across** the road.

2. **Time** (after, at, before, by, for, from etc.).

   He will return the goods **at** an early date.

3. **Agency (Instrumentality)** - (at, by, through etc.).

   Please send the parcel **by** post.

4. **Manner** (by - with)

   He faught **with** courage.

5. **Cause, reason, purpose** (of, for, from etc.).

   He did it **for** our good.

6. **Possession** (of, by, with etc.)

   He is a man **of** means.

7. **Measure, standard, rate, value** (at, by etc.).

   Bank will charge interest **at** ten per cent.

8. **Contrast, concession** (after, with etc)

   After every effort he failed.

9. **Inference, Motive, Source,** Origin (from)

   Light comes from the sun.

EXERCISE – I

Fill in the blanks with appropriate prepositions and rewrite the sentences.

1. We talked .... books.
2. The sky is .... our heads.
3. His behaviour is .... suspicious.
4. Day .... day we worked hard.
5. He fell ill on Monday and recovered four days ....
6. We shall see you .... dinner.
7. One should guard .... cold.
8. They spoke .... the scheme.
9. Don't lean .... the wall.
10. I have nothing to say .... him.
11. He gazed .... the lovely scene.
12. My memory was .... fault.
13. I walked .... him.
14. I must see you .... Monday.
15. I met him the day .... yesterday.
16. The sun has gone .... the clouds.
17. He was just .... me.
18. His class is .... his control.
19. He lives .... his income.
20. I was tired .... the evening.
21. We learn .... experience.
22. They won the match .... fifty runs.
23. He cried .... help.
24. He has purchased two books .... me.
25. This is a suitable place .... study.
26. He was absent .... college yesterday.
27. What stopped you .... coming ?
28. I remained .... the home.
29. Write your report .... ink.
30. He is .... trouble.

EXERCISE – II

Use appropriate prepositions wherever required and rewrite the sentences.

1. I want to hammer this plan your head.
2. He complained to me your rudeness.
3. He was killed an accident.
4. Have you informed them your plan ?
5. Have I asked too much you ?
6. He seems to be hard hearing.
7. He is very keen going.
8. They insisted going with me.
9. Student depend their parents.
10. He resolved making an early start.
11. He is lying a couch.
12. We went a journey.
13. I have no money to spend books.

14. The island lies the coast.
15. He jumped the fence.
16. He is the point.
17. He hit the ball the boundary.
18. There is a bridge the over.
19. He jumped the lane.
20. Your advice is very useful me.

EXERCISE – III

Construct sentences containing the following expressions.

1.	Appropriate to	2.	Allowance for
3.	Abstain from	4.	Contrary to
5.	Beneficial to	6.	Worthy of
7.	Accountable to	8.	Addicted to
9.	Entitled to	10.	Incentive to
11.	Famous for	12.	Fond of
13.	Divided into	14.	Wish for
15.	Inaccessible to	16.	Liable to
17.	Confined to	18.	Adapted to
19.	Derived from	20.	Profited by

(I) Conjunctions

A conjunction is a word which merely joins together sentences, and sometimes words.

To make sentences more compact, conjunctions are used. Conjunctions are also termed as sentence connectives.

Classes of Conjunctions

1. **Co-ordinating conjunctions :**

   (Man walks and birds fly) clauses of equal rank.

   Example : And, but, for, or, nor, also, either-or, neither-nor.

   (i)   Cummulative (adds)

   They carved not a line **and** they raised not a stone.

   (ii)   Adversative (contrast)

   He was slow, **but** he was sure.

   (iii)   Disjunctive or Alternative (choice)

   Run quickly, **else** you will not overtake him.

   (iv)   Illative (Inference)

   All precautions must have been neglected, **for** the disease spread rapidly.

2. **Subordinating Conjunctions :**

   They ran away because they were afraid.

   Example : After, because, if, that, though, although, till, before, unless, as, when, where, while.

   (i)   Time (before, till, since etc).

   He returned home **after** I left.

(ii) Cause or Reason (because, since, as)
  Since I wish it, it will be done.
(iii) Purpose (that, lest etc).
  We eat **that** we may exist.
(iv) Result (that)
  I was so tired **that** I could hardly stand.
(v) Condition (if, unless etc).
  I will come **if** you accompany.
(vi) Concession (though, although)
  Though he blamed me, yet l like him.
(viii) Comparison (than)
  Ram is stronger **than** Hari.

EXERCISE

Use the following conjunctions and rewrite the sentences given below (Unless, since, if, till, before, after, than)

1. I will not succeed .... I work hard.
2. He waited .... the train arrived.
3. Do not go out .... I leave.
4. .... you say so, they must .... believe.
5. You will get the class .... you deserve it.
6. I will be unhappy .... you do that
7. She was sorry .... she had done it.
8. He is stronger .... I am.
9. His father died .... he was born.
10. Catch him .... you can.

Uses of Some Important Conjunctions

SINCE

1. When 'since' is used as a conjunction, it gives the meaning as 'from and after the time' mentioned.

  For example : Many changes have taken place since I left Polytechnic.

2. 'Since' also gives the idea of 'seeing that', in as much as, for the reason etc.

  For example :

  Since you wish it, it will be performed.

  Since you will not work, you shall not be paid.

  Since that is the case, we shall excuse you.

OR

1. When 'OR' is used as a conjunction, it gives the meaning as 'alternative'.

  For example : One must work or starve.

2. 'OR' also gives a choice between any two in the series.

  For example : I may study Law or Engineering or Medicine or I may enter into a business.

3. 'OR' is also used to introduce alternative names or synonyms.

  For example : The violin or fiddle has become an instrument of the modern era.

4. 'OR' also gives the meaning as 'otherwise'.

  For example : We must work hard or face the worst situation.

IF

1. 'If' gives the meaning as 'on the condition'.
   For example : If that is so, I am satisfied.
2. 'If' gives the meaning as 'admitting that'.
   For example : If we are poor, yet we are honest.
3. 'If' gives the meaning as 'whether'.
   For example : I asked him if he would come.
4. 'If' gives the meaning as 'whenever'.
   For example : If we feel any doubt we enquire.

THAT

1. 'That' can be used to express reason or cause and is equivalent to because, for that, in that etc.
   For example : He was angry that he was opposed.
2. 'That' can be used to express a consequence, result etc.
   For example : I am so tired that I could not stand.

THAN

When the word 'than' is used as a conjunction, it follows adjectives and adverbs in comparative.

For example : Wisdom is better **than** riches.

I know him better **than** you do.

We will rather suffer **than** you give what you want.

LEST

It can be used as a subordinating conjunction expressing a negative purpose, and is equivalent to 'in order that .... not'. For fear that etc.

For example : They fled lest they should be punished.

WHILE

While is used to mean :
(i) During the time that, as long as
(ii) At the same time that
(iii) Whereas, as

For example : **While** there is life, there is hope.

The girls sang **while** the boys played.

While it is true of some, it is not true of all.

ONLY

The word 'only' can work as a conjunction meaning except that, but, were it not (that),

For example : They do well, **only** that they are nervous at the start.

I would agree, **only** in difficulties.

BECAUSE, FOR, SINCE

Of these three conjunctions, 'because' denotes the closest causal conjunction; 'for' the weakest and 'since' comes between the two.

EXERCISE

Use the following conjunctions in sentences of your own.

1. Than 2. Since 3. Except 4. Only 5. That.

(J) Interjections

An interjection is a word which expresses sudden feeling or emotion.

For example :   Alas ! — (expresses grief)

                      Ha ! (suprise etc).

For example :   Alas ! he is no more.

Hurrah ! we win the match. (express 'joy').

In the above sentences, the words 'alas', 'ha', 'hurrah', etc. have no definite meaning of their own. They are used to express feelings or emotions alone. They are used for giving emphasis on the corresponding feelings.

(K) Punctuation

Punctuation :

The modern trend is to use punctuation marks, only where they are required, in order to keep the clarity of expression.

Punctuation Marks :

1.   Comma   [,]
2.   Semicolon   [;]
3.   Colon   [:]
4.   Full stop   [.]
5.   Question mark   [?]
6.   Quotation marks   [" "]
7.   Dash   [—]
8.   Hyphen   [–]
9.   Brackets   [( )]

Capitalisation :

Sometimes capitals are used to make a written message easy to read and understand.

Generally, capitals are used for names of places, persons, organisations, designations, days, months, holidays, religious days, abbreviations etc.

Numerals :

Numerals are used while writing messages to make the content easy to understand.

For example :

1.   I, II, III, IV
2.   10 hours
3.   Sixty-two motors
4.   Six grams of salt
5.   Rs. 3/-
6.   Fig. 10

Abbreviations :

Abbreviations are the short forms of words or phrases :

For example,

1.   Sun   –   Sunday
2.   Wed.   –   Wednesday
3.   Jan.   –   January
4.   Smt.   –   Shrimati

5. Prof. – Professor
6. amp. – amphere
7. g – gram
8. km. – kilometer
9. do – ditto
10. p – page
11. mm – millimeter/millimeters.

Certain Spelling Rules

I. A silent 'e' at the end of the verb is deleted while adding 'ing'

For example,

1. Write – Writing
2. Make – Making
3. Hope – Hoping

II. A single 'e' at the end of a word is deleted when suffixed by 'able' – except where 'e' is needed to soften 'c' or 'g'.

For example :

1. love – lov<u>able</u>
2. debate – debat<u>able</u>
3. notice – notic<u>eable</u>
4. manage – Manag<u>eable</u>

III. In case combination 'le.', or 'el,' is pronounced as 'ee', 'l' is used before 'e' except after 'c' – believe, field, grief, relief.

For example :

1. Receive
2. Ceiling
3. Perceive

Exception Seize, Species, Seige

IV. If words end with 'l' preceded by only one vowel, the last 'L' is doubled when added by a suffix.

Jewel – Jeweller

Quarrel – Quarrelled

V. Words ending in 'y' preceded by a consonant change to the plural following the rule : <u>Change the 'y' into 'i' and add 'es'.</u>

1. Dry – Dried
2. Merry – Merrily
3. Ally – Alliance
4. Joy – Joyous
5. Boy – Boyish

Words ending in 'y' preceded by a consonant change to the plural following the below rule :

Change the 'y' into 'i' and add 'es' :

1. Army – Armies
2. Copy – Copies
3. City – Cities

But if the final 'y' is preceded by a vowel, just add 's'

1. Monkey   –   Monkeys
2. Joy   –   Joys
3. Day   –   Days

The same rule is applied when the tense of a verb changes :

1. Reply   –   replied
2. Apply   –   Applied
3. Play   –   played
4. Destroy–   Destroyed

Exceptions :

1. pay   –   paid
2. lay   –   laid
3. say   –   said

VI. Words ending in a single consonant preceded by a stressed vowel, double the final consonant when adding ed, ing, er.

1. Thin   –   Thinner
2. Run   –   Running
3. begin   –   beginning
4. bat   –   batted

In case the vowel is not stressed, the consonant is not doubled.

1. benefit –   benefited
2. Profit   –   profiting

VII. The following words are written as one word :

Almost   –   meanwhile
anyone   –   newspaper
anything   –   otherwise
already   –   sometimes
anyhow   –   something
afterwards   –   outside
cannot   –   whenever
everyday   –   throughout
however   –   within
into   –   without
moreover

VIII. The following words are written separately

all right   –   all round
inspite of   –   per cent
do not   –   young man
in fact

Spelling :

While writing a message it is always necessary to give full attention towards correct spelling of words, otherwise it creates confusion in comprehending the message and leads to bad impression about the writer.

Certain commonly misspelt words

Abbreviate	acceptance	absence	accessory
accede	accommodate	accessible	accumulate
accidentally	achievement	accountant	acquaintance
accustomed	actually	acknowledge	advantageous
acquire	administrator	adherence	affectionately
advertisement	advisable	advisory	aggrieved
agreeable	affiliate	agenda	ambiguity
ambiguous	alignment	allowance	annually
apologise	analyse	analytical	appropriate
approximate	appearance	appendix	audience
artificial	arbitrate	argument	auxiliary
auditor	ascertain	assessor	
average	authorise	authoritative	
abridgement	available	abundance	
bankruptcy	beggar	beginning	bureaucracy
believe	beneficent	behaviour	
brochure	budget	benefitted	
Calender	cancellation	capacity	collateral
characteristic	casually	catalogue	commission
colleague	clearance	clientele	comparison
comprehensive	commencement	comparable	conscience
conscientious	commitment	conference	consideration
consignor	conscious	consecutive	contradict
controllable	controversial	convene	courteous
convertible	co-operation	career	
creditor	curiosity	changeable	
debenture	discipline	deference	definitely
deductible	demonstrate	demurrage	dependence
deliberate	development	dissent	disapprove
determine	deceive	decision	distinguish
earnest	expiry	explain	explanation
eighth	economical	economic	efficient
eliminate	embrassed	elementary	eligible
envelope	equity	endeavour	experience
excellent	exception	excise	extraordinary
exclamation	exhibition	existence	
familiar	fitted	Fourteen	fulfilled
feasible	forty	foreign	fulfilment
February	forecast	forfeit	
generally	grandeur	guarantee	
genuine	grievance	guardian	
haphazard	height	honorarium	hypothecation
harass	hindrance	humour	
hedging	honorary	humorous	

illiterate	irrevocable	incorporate	incorrigible
incredible	immediately	indemnity	indispensable
industrialisation	indemnify	inferior	influence
influential	inference	initial	instalment
inspection	instruction	instrument	interrupt
interview	invoice	irrelevant	irreparable
January	jealous	judgement	jurisdiction
knot	knowledgeable		
labelled	labour	language	lawyer
leather	legible	liaison	liberalise
lien	liquidate	liquor	literature
lodging	lottery	luxury	luxurious
machinery	manageable	mediate	misappropriate
magnificent	manufacture	memorandum	messets
maintain	measure	millionaire	miscellaneous
maintenance	mechanism	messenger	
necessary	neither	neuter	nineteen
neighbour	ninety	ninth	northern
notable	noticeable	November	numerous
necessarily	negligible	negotiate	
obedience	October	offence	official
occurrence	omission	omitted	opinion
opportunity	opposition	original	overwhelming
objectionable	obvious	occasionally	
pamphlet	patronage	permitted	perpetual
partner	percentage	persuasion	petition
perceive	permissible	predecessor	preferred
permanent	persuade	prevalent	preference
persistence	possibility	proceed	prior
pioneer	preliminary	programme	procedure
prejudice	probability	prospectus	provision
privilege	psychology	pursue	pursuit
profession	parliament	particular	
propriety	peculiar	penalty	
parallel	performance	perishable	
quarrelled	questionnaire	queue	quorum
receipt	recurrence	regrettable	repetition
rectify	registration	religion	restaurant
register	relieve	recommend	revocable
reliable	repeat	referred	reputable
resolution	resource	reference	reveal
revenue	review	relevant	ridiculous
receive	recognise	remembrance	

salable	seize	summary	straight
sampling	separate	sincerely	subscribe
satisfactory	simultaneous	skill	superintendent
Saturday	specimen	skilful	superior
schedule	succeed	speculation	supervision
scientific	sufficiently	statistics	surely
secretary	suitable	statute	
tariff	technical	technique	temperament
temporary	tenure	thoroughly	tiresome
transferred	treasurer	truly	twelfth
unanimous	underpaid	utilise	utterance
until	usually	unique	
vacation	vacuum	valuable	variable
variety	vehicle	voluntary	volunteer
waive	warranty	willing	
withdrawal	writing		

4. CORRECT THE ERRORS FROM THE SENTENCES

(Correct Usage)

Following are the important hints which are useful for correcting the incorrect sentences.

(I) Agreement of the verb with the subject :

(a) A verb must agree with its subject in Number and Person.

For example :

She is at home.

The number of women was small.

(b) The noun following *one of the* best should be in plural number.

For example :

Utkarsh is one of the best students in the group.

(c) Two or more singular nouns or pronouns joined by *and* require a plural verb.

For example :

– Sometimes *luck* and power *have* no relation.

(d) If the nouns denote the *same person* or *thing*, the verb and the pronoun should be singular.

For example :

Slow and steady *wins* the race.

The writer and poet *is* no more.

Time and tide *waits* for no man.

(e) In case the two nouns connected by *and* have an article placed before each of them, they refer to two different persons or things, and take a verb in the plural.

The writer and *the philosopher* are no more.

(f) When the subject consists of two or more nouns joined by *with, together with, or as well as,* the verb agrees with the noun preceding these.

For example :

English, *as well as* Marathi, *is* taught in the class.

The politician *with* his wife and the rest of the people, *was* present.

(g) Two or more singular subjects connected by *or* or *nor* require a singular verb.

For example :

Our happiness or our luck is largely due to our own deeds.

Neither youth *nor* beauty *is* safe against age.

(h) When two or more subjects of different persons are joined by *or,* or *nor,* the verb agrees with the one nearest to it.

For example :

Either you or he <u>is</u> wrong.

(i) The distributive *each, every, either, neither,* refer to each individual of a class and are therefore followed by a verb in the singular.

For example :

Every one of the *robbers was* arrested.

(j) When two nouns are qualified by *each* or *every* though connected by *and,* they require a singular verb.

For example :

Each day and each hour brings its duties.

(k) Some nouns which are singular in form, but plural in use, take a plural verb.

For example :

Cattle are coming.

Mankind are liable to make mistakes.

(l) Conversely, some nouns are in plural uniform but singular in meaning, these require the verb in singular.

For example :

This news *is* so good that it cannot be true.

The reward of sin is death.

(m) A collective noun requires a verb in the singular when it denotes the collection as a whole, and the verb in the plural when it denotes the individual in the collection separately.

For example :

The group was dismissed.

The whole group were in tears.

(II) Pronouns :

(a) When the complement of the verb to be is a pronoun, it is placed in the nominative case.

For example :

It is I *who* am your best friend.

It is he *who* did it.

(b) When the object of a verb or of a preposition is a pronoun, it is placed with the objective case.

For example :
Between you and me (not I) he is a rogue.

(c) A pronoun must agree with its Antecedent in Gender, Number and person.

For example :
Every person must sustain his own burden.
Each of the boys went to his house.

(d) The indefinite pronoun one must always be followed by one, one's. For example,
One must be definite when *one* makes the statements.
One must do one's job.

(e) The case of the Relative Pronoun. The Relative Pronoun is in the Nominative Case, when it is the subject of a verb, and in the objective, when it is the object of a verb, or preposition.

For example :
The prize was given to the student who, the masters said really deserved it.
The boy, whom you praised so much has failed.

(f) A Relative Pronoun agrees with its Antecedent in Gender, Number and person.

For example :
He is one of those men who are always finding fault with other people.

(g) When the verb is understood, as is always used.

For example :
This is the same notebook as mine.

(III) Adjectives :

(a) The comparative degree is used in comparing two things or classes of things.

For example :
Metal is more useful than wood.

(b) The superlative degree is used in comparing one thing with all others of the same kind.

For example :
Ramesh is the tallest boy in the group.

(c) When the comparative degree is used, the latter term of comparison should exclude the former.

For example :
Aditee is greater than any other novelist.

(d) When the superlative is used, the latter term should include the former.

For example :
He was the best person in the group.

(e) When the second term of a comparison is given, it must correspond in construction with the first.

For example :
The progress of 'A' is greater than that of 'B'.

(f) Double comparative or superlatives should not be used.

For example :
I am happier than he.
It was the unkindest cut of all.

(g) The words, interior, junior, posterior, prior, senior, superior, take *"to"* instead of *"than"*.

For example :

He is junior to me.

(h) Preferable has the force of a comparative and is followed by *"to"*.

For example :

Health is preferable to wealth.

(i) The following adjectives do not admit of comparison : unique, perfect, universal, extreme, entire, square, round, full, chief, complete, ideal.

For example :

This is a unique example.

(j) The Plural forms *these* and *those* are often wrongly used with *kind* and *sort* which are nouns in the singular.

For example :

Incorrect : I never read those kind of a novel.

Correct : I never read *that kind* of novels.

Incorrect : Where can I buy *these sort* of paper ?

Correct : Where can I buy *this sort* of paper ?

(IV) Articles :

(a) The Indefinite Article *'A'* is used.

    (i) Before nouns beginning with a consonant.

        For example :

        a boy, a girl, a hero, a woman, a year.

    (ii) Before "U" pronounced as "yoo"; as a University; a useful book, a European, a Unit.

    (iii) Before "O" pronounced as *"WU"*; as a one-eyed giant.

        Such a one.

(b) *'AN'* is used.

    (i) Before words beginning with a vowel, as an owl.

        For example :

        an elephant

        an umbrella

    (ii) Before a silent "h".

        For example :

        an hour

        an heir

    (iii) Before words beginning with "h" and not accented on the first syllable, an is used.

        For example :

        an hotel

        an historical book

(c) The Definite Article "The" is used.

    (i) To point out a particular person or thing.

        For example :

        This is the book of which I spoke.

    (ii)   Before singular noun meant to represent a whole class.

        For example :

            The rose is the queen of flowers.

    (iii)   With names of seas, rivers, gulfs, oceans, groups of island, and mountain ranges.

        For example :

            The Ganges is the holiest river in India.

            Pune is on the Mula and Mutha river.

    (iv)   Before well-known single objects.

        For example :

            The sun

            The moon

            The sky

    (v)   With superlatives.

        For example :

            The darkest cloud has a silver lining.

    (vi)   Before an adjective used as a noun.

        For example :

            The poor are happier than the rich.

    (vii)   As an adverb with comparative.

        For example :

            The more man has, the less happy he is.

  (d)   The article is *omitted.*

    (i)   Before a noun used in a general sense.

        For example :

            Man is mortal.

    (ii)   Before proper nouns.

        For example :

            Mumbai is a city of tall buildings.

    (iii)   Before material nouns.

        For example :

            Gold is a precious metal.

    (iv)   Before abstract nouns.

        For example :

            Beauty needs no ornament.

    (v)   Before the noun following 'kind of'.

        For example :

            What *kind* of flower is it ?

(V) Verbs :

  (a)   A verb should be made to agree with the subject and not with the complement.

    For example :

        A special character of the exhibition was the workshops.

  (b)   In a compound sentence, a single verb cannot be made to do duty for two subjects if they are not in same number.

    For example :

    Incorrect : The leader was hanged and his followers *arrested.*

    Correct : The leader was hanged and his followers *were arrested.*

(c)   The conjugation of the verb *lay* and *lie* should be carefully noted. The verb lay is transitive and is always followed by an object; the verb *lie* is intransitive and cannot have an object.

> *lay, laid, laid.*
>
> > Lay the note book on the table.
> >
> > She laid it on the table.
> >
> > She has laid it on the table.
>
> *Lie, lay, lain.*
>
> > He lay on the sofa.
> >
> > Let him lie there.
> >
> > He has lain there for hours.

(d)   **The Split Infinitive :** In English the infinitive is often preceded by *to*. The "split" of an infinitive is to place an adverb between this to and the rest of the verb.
Incorrect : It is essential to thoroughly examine the question.
Correct : It is essential to examine the question thoroughly.

(e)   **The Unrelated Participle :** As the participle is an adjective, there must be some noun that it limits. If this noun is omitted we have the error of the unrelated participle.
Incorrect : Arriving at the house, the door was shut.
Correct : When he arrived at the house, the door was shut.

(f)   The pronoun governing a Gerund should be in the possessive case.
For example :

> What is the use of my going there ?

(g)   The noun governing a Gerund should also be put in the possessive case with 'S, if it is of such a kind to take that infection.
For example :

> I am glad that Sachin's coming here.

(VI) Uses of Shall and Will :

(a)   Shall is used in the first person to express simple future time.
For example :

> I shall be highly obliged.

(b)   When shall is used in the second or third person, it usually expresses :

(i)   A command, as

> You shall not take.

(ii)   A promise, as

> You shall have a gift tomorrow.

(iii)   A threat, as

> You shall be punished for this act.

(iv)   Determination, as

> The time will come when you shall hear me.

(c) Will is used in the second and third persons, to express simple future time.

For example :

You will meet her on the road.

(d) Will is used in the first person to express

(i) Willingness

We will do that work.

(ii) A promise

I will give you my note book.

(iii) A threat

I will punish you if you do not work.

(iv) Determination

I will succeed or die in the attempt.

(e) (i) Shall is used in asking question in the person.

Shall I ring the bell ? (The action is dependent on the will of the person addressed).

(ii) Either shall or will is used, according to the answer expected, in asking questions in the second and third persons, as,

Shall you sign the agreement (I shall sign the agreement).

Will you give me the vehicle ? (I will give you the vehicle).

(VII) Adverbs, Prepositions and Conjunctions :

(a) An adverb should be so placed that there can be no doubt as to its relation to the rest of the sentence.

For example :

I was greatly surprised at the result.

(b) Two negatives should not be used in the same sentence, unless we wish to make an affirmation. For example,

I can do nothing.

(c) The use of never for not is incorrect.

For example :

Incorrect : He was never born in India.

Correct : He was not born in India.

(d) *At* is used before the name of a city or a town; *in* is used before the names of countries.

For example :

He was born *at* Shivaji Nagar *in* Pune.

(e) *By* refers to the agent; *with* to the instrument as.

For example :

This book was written *by* me *with* a ball pen.

(f) *In* is used in speaking of things at rest; into is used in speaking of things in motion.

For example :

She was sitting *in* the room.

She walked *into* my room.

(g)   The adverb *like* is often wrongly used as a conjunction instead of as,

He speaks *as* his father does.

(h)   Scarcely should be followed by *when*; as,

Scarcely had she gone out, *when* her friend arrived.

(i)   *No sooner* must be followed by *than*;

For example :

No sooner had the bus arrived *than* she stepped out.

(j)   In contracted sentences conjunctions are often wrongly omitted after *Adjectives*.

Incorrect : This is as good if not better than that.

Correct : This is a good as, if not better than that.

Common Errors Corrected

1.   Incorrect : Neither he nor I are in the wrong position.

Correct : Neither he nor I am in the wrong position.

2.   Incorrect : The fleet were under orders to move.

Correct : The fleet was under orders to move.

3.   Incorrect : He, with his brother, was among the first to come.

Correct : He, with his brother, were among the first to come.

4.   Incorrect : Nothing but grave and serious studies delight her.

Correct : Nothing but grave and serious studies delights her.

5.   Incorrect : He is taller than me.

Correct : He is taller than I.

6.   Incorrect : I thought it was him.

Correct : I thought it was he.

7.   Incorrect : Suraj is a man whom I think deserves praise.

Correct : Suraj is a man who I think deserves praise.

8.   Incorrect : Nobody in their senses would have done such act.

Correct : Nobody in his senses would have done such an act.

9.   Incorrect : Of the two I think her to be the best.

Correct : Of the two I think her to be better.

10.   Incorrect : I do not like those kind of books.

Correct : I do not like those kinds of books.

11.   Incorrect : He is stronger than any person living.

Correct : He is stronger than any other person living.

12.   Incorrect : Anirudha was the wisest of all other men.

Correct : Anirudha was the wisest of all men.

13. Incorrect : This is more preferable than that.

    Correct : This is preferable to that.

14. Incorrect : The Secretary and the treasurer is absent.

    Correct : The Secretary and the treasurer are absent.

15. Incorrect : Yashwant was the orator and the statesman of his period.

    Correct : Yashwant was the orator and statesman of his period.

16. Incorrect : Draw map of India.

    Correct : Draw a map of India.

17. Incorrect : Time makes worst enemies friends.

    Correct : Time makes the worst enemies friends.

18. Incorrect : They never fail who die in great cause.

    Correct : They never fail who die in a great cause.

19. Incorrect : It is a week since the holidays commenced.

    Correct : It was a week since the holidays commenced.

20. Incorrect : My father has been ill since three days.

    Correct : My father has been ill for the last three days.

21. Incorrect : You will bear it as you have so many things.

    Correct : You will bear it as you have borne so many things.

22. Incorrect : He not only lost his card, but also his mobile.

    Correct : He lost not only his card, but also his mobile.

23. Incorrect : Please except my best wishes.

    Correct : Please accept my best wishes.

24. Incorrect : What is the use of him coming here ?
    Correct : What is the use of his coming here ?

25. Incorrect : Walking in the field, the gate suddenly opened.
    Correct : While we were walking in the field, the gate suddenly opened.

26. Incorrect : I will be drowned and nobody shall save me.
    Correct : I shall be drowned and nobody will save me.

27. Incorrect : Do it like I do.

    Correct : Do it as I do.

28. Incorrect : Scarcely had he gone, than a gentleman knocked at the door.
    Correct : Scarcely had he gone, when a gentleman knocked at the door.

29. Incorrect : This is as good if not better than that.
    Correct : This is as good as that, if not better.

30. Incorrect : They only work when they have no money.
    Correct : They work only when they have no money.

31. Incorrect : He saw countless numbers of the dead riding across the field of battle.

    Correct : Riding across the field of battle, he saw countless number of dead.

32. Incorrect : For sale, piano, the property of a musician with carved parts.

Correct : For sale, a piano with carved parts, the property of a musician.

33. Incorrect : The hall is full, there is no place for any more.

Correct : The hall is full, there is no room for any more.

34. Incorrect : Good night, Sir, I am glad that you have come.

Correct : Good evening, Sir, I am glad that you have some.

35. Incorrect : She came by the 8.30 o' clock train.

Correct : She came by the 8.30 train.

36. Incorrect : Her fault is such as cannot be pardoned.

Correct : Her fault is such which cannot be pardoned.

37. Incorrect : She was prevented to go out.

Correct : She was prevented from going out.

38. Incorrect : No sooner she had returned then she went out.

Correct : No sooner had she returned than she went out.

39. Incorrect : It is much too hot today.

Correct : It is very hot today.

40. Incorrect : Every boys have their notebooks.

Correct : Every boy has his notebooks.

41. Incorrect : One must obey to his parents.

Correct : One must obey one's parents.

42. Incorrect : When trying to swim, the mouth must be kept above water.

Correct : When one is trying to swim, one must keep one's mouth above water.

43. Incorrect : He failed due to his weakness in Marathi language.

Correct : He failed owing to his weakness in Marathi language.

44. Incorrect : Ambition is one of those acts that is never satisfied.

Correct : Ambition is one of those acts that are never satisfied.

45. Incorrect : All his income is spent and all his hopes ruined.

Correct : All his income is spent and all his hopes are ruined.

46. Incorrect : He is one of the richest, if not the richest man in India.

Correct : He is one of the richest men, if not the richest in India.

47. Incorrect : This rule may and ought to be disregarded.

Correct : This rule may be and ought to be disregarded.

48. Incorrect : He sold many paper and pens.

Correct : He sold much paper and many pens.

49. Incorrect : A refugee is a person who keeps order at a football match.

Correct : A referee is a person who keeps order at a football match.

50. Incorrect : I cannot help but think that you are correct.

Correct : I cannot help thinking that you are correct.

51. Incorrect : My house is not larger than many others have built for themselves.

Correct : My house is not larger than those that many others have built for themselves.

52. Incorrect : He sent them his compliments.

Correct : He sent them his complements.

53. Incorrect : The rice on this land is finer than last year.

Correct : The rice on this land is finer than it was last year.

54. Incorrect : He gave the table to the poor man that had a broken leg.

Correct : He gave the table that had a broken leg to the poor man.

55. Incorrect : They builds a theatre to accommodate four hundred people ninety feet long.

Correct : They built a theatre ninety feet long to accommodate four hundred people.

56. Incorrect : I showed him the cold shoulder.

Correct : I gave him the cold shoulder.

57. Incorrect : She is more educated than me.

Correct : She is more educated than I.

58. Incorrect : When looking through a mist, things appear unnaturally large.

Correct : When seen through a mist, things appear unnaturally large.

59. Incorrect : I saw through the window some girls going to school.

Correct : I saw some girls going to school through the window.

60. Incorrect : He was not only the maker of a nation but of a language.

Correct : He was the maker not only of a nation, but also of a language.

61. Incorrect : She never remember to have met her equal.

Correct : She does not remember to have ever met her equal.

62. Incorrect : The account of the murder made his flesh creep.

Correct : The account of the murder made his blood creep.

63. Incorrect : Being interested in fishes, an aquarium is always to be found in their house.

Correct : They are so interested in fishes, that an aquarium is always found in their house.

64. Incorrect : Running up against my friend in the park, she told me the latest news.

Correct : Running up against my friend in the park, I heard from her the latest news.

65. Incorrect : While digging the foundations for a building, a piece of Maurya wall was discovered.

Correct : While the workmen were digging the foundations for a building, a piece of the Maury wall was discovered.

66. Incorrect : Trying to escape, the dog met her at the gate.

Correct : As he was trying to escape, the dog met her at the gate.

67. Incorrect : Having finished the chapter, the volume was shut.

Correct : Having finished the chapter, she shut the volume.

68. Incorrect : One should not boast of her success.

Correct : One should not boast of one's success.

69. Incorrect : They promised to promptly pay the sum.

Correct : They promised to pay the sum promptly.

70. Incorrect : They are more stupid and inferior to their friends.

Correct : They are more stupid and inferior than their friends.

71. Incorrect : Each of the boys went to their separate rooms to rest and calm themselves.

Correct : Each of the boys went to his separate room to rest and calm himself.

72. Incorrect : She told me and you that she will come.

Correct : She told you and me that she would come.

73. Incorrect : This was her who they knew.

Correct : This was she whom they knew.

74. Incorrect : She is a practicable person.

Correct : She is a practical person.

75. Incorrect : That building is not advertised for sale but for hire.

Correct : That building is advertised not for sale but for hire.

76. Incorrect : A nurse maid is wanted for a baby about twenty five years old.

Correct : A nurse maid about twenty five years old is wanted for a baby.

77. Incorrect : She has taken a leap into the dark.

Correct : She has taken a leap in the dark.

78. Incorrect : Whom did they say was the oldest of the two ?

Correct : Who did they say was the older of the two ?

79. Incorrect : I am confident to win the prize.

Correct : I am confident of winning the prize.

80. Incorrect : The three rivers are connected to one another by a canal.

Correct : The three rivers are connected with each other by a canal.

81. Incorrect : Since several days she was been ill from fever.

Correct : For several days she has been ill of fever.

82. Incorrect : He left the hotel where he had been staying in a motor car.

Correct : He left in a motor car the hotel where he had been staying.

83. Incorrect : Between you and I, neither of us were right.

Correct : Between you and me, neither of us was right.

84. Incorrect : She was laying stretched on the ground.

Correct : She was lying stretched on the ground.

85. Incorrect : My mother and myself are going into the city.

Correct : My mother and I are going into the city.

86. Incorrect : She was a clever, or even cleverer than her sister.

Correct : She was a clever as, or even cleverer than her sister.

87. Incorrect : We are very annoyed to hear it.

Correct : We are much annoyed to hear it.

88. Incorrect : These flowers smell sweetly.

Correct : These flowers smell sweet.

89. Incorrect : She is fare, but week in health.

Correct : She is fair, but weak in health.

EXERCISE

(A) Fill in the blanks with a verb in agreement with its subject.

1. Curry and rice ……….. my favourite food.

2. No news ……….. good news.

3. A large number of men ……….. present.

4. Each of the suspected men ……….. present.

5. Every town and village ……….. burnt.

6. Neither your friend nor you ……….. present.

(B) Fill in the blanks with "who" or "whom" :

1. I am he ……….. you seek.

2. We do not know ……….. will go.

3. ……….. does he wish to see ?

4. ……….. is it that you wish to see ?

5. It was the President ……….. we asked to speak.

6. Do you know ……….. I am ?

7. ……….. do you take me to be ?

8. Where is the man about ……….. he was speaking ?

(C) Rewrite the following sentences, inserting the preferred form :

1. He is the (wiser, wisest) of the two.

2. This book is the (better, best) of the three.

3. The crocodile is larger than (any, any other) reptile.

4. The elephant is the largest of (all, all other) animals.

5. This tree has a wider spread than (any, any other) tree.

6. The "Times of India" has the (larger, largest) circulation of all newspapers.

7. Sunil plays better cricket than (any, any other) boy.

(D) Fill up the blanks with 'a' or 'an' or 'the' :

1. She admires ……….. beautiful.

2. Be ……….. hero in the strife.

3. Such ……….. one is seldom found.

4. Hindi is ……….. language of ……….. people of India.

5. ……….. University will shortly be established here.

(E) Insert "shall" or "Will" in the blanks :

1. I ……….. punish you if you do that again.

2. He ……….. control his temper.

3. I promise that you ……….. have the money.

4. You ……….. do your duty, if you ……….. shine in life.

Part – III

PARAGRAPH WRITING

(1) Types of Paragraphs (Narrative, Descriptive, Technical)
(2) Unseen Passage for Comprehension

[Marks 08]

(1) PARAGRAPH WRITING

Introduction

The basic object of paragraph writing is to communicate. It is the first step in developing an ability to write a piece of composition. Practice of writing paragraph gives an opportunity to the students to develop and improve the power of expression. People engaged in the technical field are no exception to this rule. They are constantly required to communicate to others in the form of technical notes, technical reports, explaining certain technical processes, functions of tools and machines, preparation of work projects etc.

1 **Unity of thought :** Although there is no precise definition of paragraph writing, it is generally considered as a group of sentences which deal with one unit of thought forming a definite state in the development of the subject or the theme. Paragraph writing, technical or general as the case may be, involves two important points viz.

(a)   The paragraph deals with only one topic; and

(b)   The topic is a significant and important part of the subject or theme as a whole.

We should be aware that in writing we are concerned with different units of thoughts which vary in the degree of their importance. In fact, the theme or subject is itself a unit of thought. However, it consists of different paragraphs bringing out the important features of different aspects of the theme. Similarly, each paragraph is also a group of sentences, each one dealing with a still smaller division of thought. The paragraph therefore, differs from the sentences in that it covers a larger topic that carries discussion forward through a definite stage.

2 **The Topic Sentence :** While starting to write a paragraph, the topic of the paragraph is often started in a brief sentence, called the 'topic sentence' which sums up the entire contents of the paragraph. The topic sentence is generally taken in the beginning. However, there is no hard and fast rule about its place. Sometimes, it can also come at the end or may find place in the middle of paragraph. For the benefit of the students it is suggested that the topic sentence is taken in the beginning so that there is no possibility of wandering away from the main subject. The topic sentence should give exact statement of the material it covers. The technical paragraph must contain discussion of the subject stated in the topic sentence.

3 **Logical order :** As we have seen that a paragraph is a group of sentence, it follows that every sentence in a paragraph must bear relevance to the topic and must contribute something worthwhile to its development. Thus, the thought is presented as a unit which can be readily

(3.1)

understood by the reader. This is possible only if development of a paragraph is done in a logical order, where one thought leads to another so that the flow of the thought is kept unbroken. For achieving coherence or continuity of thought, certain conjunctions such as hence, but, and, then, therefore, so, are very useful. They establish relationship between the two words or sentences brought together. This does not, however, mean that there should be no variations in the sentence patterns. For bringing variety, liveliness and making paragraph interesting, the sentence pattern must vary. Technical students must try to write simple structural patterns and achieve as much variety as possible.

4 **Length of paragraph :** As a general principle, students should avoid frequent use of short paragraphs, each consisting of only a few lines. This leads to the possibility of dividing a significant topic into two or more groups each one dealing with comparatively unimportant details. As a result, the pieces appear to be a series of jottings or notes rather than a well-organised discussion on significant topic. Such small paragraphs make it difficult for the readers to properly understand the development in the topic. This may also lead to another flaw in the paragraph. The topics of sufficient importance may be neglected and not given proper justice in its development. Hence the technical students should try to combine and bring together into one paragraph all those groups of sentences which logically belong together and appropriately amplify them. This will also help to maintain a logical sequence of the development of theme or subject.

5 **Ending the Paragraph :** The end of the paragraph, like the end of the whole article, should give the impression of completeness. It is sometimes secured by restatement or paraphrasing of the topic. Such an ending makes a good impression of the subject of the paragraph and keeps its impact on the minds of the readers for a longer period. Hence, the concluding sentences of a paragraph is as important as the opening sentence. The first sentence should arouse the reader's interest and the last sentence should satisfy it. The last sentence should sum up the contents of the paragraph or contain some brief comment on what has gone before. It must make the reader feel that the paragraph is closed.

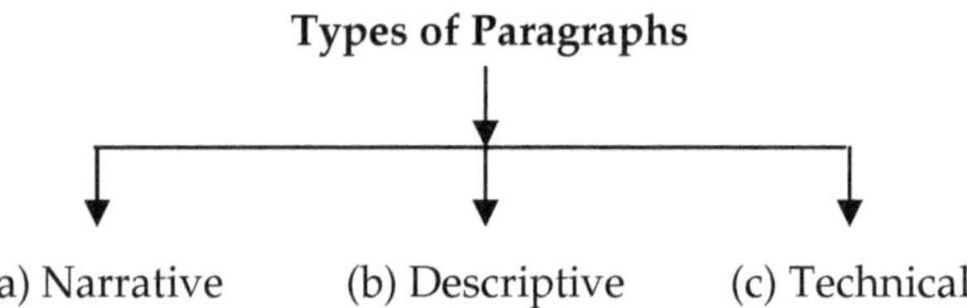

The following are the different types of paragraphs which are as below :

(a) **Narrative :** Narrative paragraph is one which describes an incident, a happening or an anecdote. The account should be clear and interesting. Such a paragraph should as far as possible be dramatic and full of suspense. The following devices are employed in constructing a paragraph depending upon the nature of the subject or theme one has to deal with.

(i) Similarities,	(ii) Definitions,
(iii) Comparison and Contrast,	(iv) Differences,
(v) Reasoning and	(vi) Facts and Figures.

(b) **Descriptive :** Descriptive paragraph is one where the reader may be able to get a clear visual picture of the matter in his mind. In such paragraphs, only significant or important details about the thought should be described, so that the picture would not be hazy and confused.

(c)   Technical : The perfect knowledge of technical paragraph writing is essential for an engineer. He is expected to write several paragraphs and technical matter during his career. The purpose of every writing is to give information to some person or a body of persons not in possession of the full knowledge of the facts of the subjects with which it deals.

Technical writing is a new and an emerging field. Since the inception of (I.T.) Information Technology Industry in our country, various Industrial establishment have emerged. These fields include graphic designing, networking, security management, technical writing and many more. Technical writing came into existence because of I.T. field. However, application of technical writing is not limited to I.T. field, but also includes writing skills which is used in other industrial activities.

Technical writing means conveying technical information to various groups of people who are interested and aspiring technical information in an appropriate manner and in a way to understand and may include printed documents as well.

(d)   Technical Paragraphs : Different changes of technical writing.

(i)   A person who is having required technical knowledge and an appropriate understanding can become technical writer.

(ii)   It is not necessary to have a degree in Arts or engineering to be a technical writer. However, one should have the ability to grasp the subject.

(iii) Education in technical field will be helpful to understand highly technical subjects in complete manner.

(iv)   Technical writing skill will help a person to get an appropriate skilled job in industry and financial institutions.

Essentials of Technical Paragraph Writing :

(1)   A Writer of the technical report must *possess sufficient knowledge* of the matter to be reported.

(2)   An Engineer is expected to *write technical report* whenever it is required.

(3)   Technical person has to prepare reports on *multifarious activities* of the organisation after detailed investigation.

(4)   Every Engineer is an executive position should be able *to set out in proper form his opinion* about the course to be adopted in any matter connected with the sphere of his responsibility.

(5)   The subjects or which technical paragraphs are required to be written are innumerable.

(6)   It is *not possible to formulate exact rules* about writing the technical paragraphs.

(7)   *No hard and fast rules can be given.*

Following are the General Guiding Principles that can be used for Technical Paragraph Writing.

(1)   Technical paragraph writing *must be brief,* clear and convincing.

(2)   It must be informative.

(3)   It must contain *relevant information* required by the reader.

(4)   Good technical graph must *be objective.* There is no place for expression of feelings of the writer.

(5)   It must be based on *observations* and inferences.

(6)   Technical paragraph must *be factual.* It must be limited to general observation and it must aim at clarity, directness and conciseness.

(7)   Technical paragraph must help the reader to get quickly what he requires, graphs, tables and charts. It should save the time in understanding.

(8)   The language must be simple and lengthy sentences are to be avoided.

(9)   Specific and correct technical words are to be used.

(10) Standard abbreviations are used in writing technical paragraph.

Example No. 1 : On Technical Paragraph on newly purchased lathe machine.

A newly purchased Lathe machine has been installed in the Production Department section of this Company just fifteen days back. I have, however, to report that the Lathe is not operating well since last week or so. The Lathe in question was purchased by our Company Representative one month back after making due inspection at Mumbai showroom and was transported from there to our factory site. During this transit, it seems that some defects have occurred due to faulty transport handling system. The defect was noticed only at the time of its installation. Such a major defect in a newly purchased machine was never expected. I personally contacted the transport company and enquired about the transportation of Lathe. The Company has point blank accepted that there were some defects in the crane used for lifting the lathe. As a result, the complete consignment was dropped on to the ground. While lifting and placing it on the vehicle of transport. In my opinion, we must claim compensation for the damage either from the transport company or the manufacturer as per the details of the contract. I do not know whether the consignment was insured before despatch.

GENERAL HINTS FOR PARAGRAPH WRITING

1.   Select the topic.

2.   Think about the topic from all angles.

3.   Ask yourself the questions – what, why, where, when, which, whom and get certain answers.

4.   Jot down answers to these questions.

5.   Arrange them in a sequence or order.

6.   Connect them with proper conjunctions.

7.   Think about starting and ending sentences.

8.   Start writing a paragraph keeping in mind the following principles :

    (a)   Unity of thought;   (b)   Statement of topic;

    (c)   Logical Order of thoughts;   (d)   Coherence;

    (e)   Arrangement;   (f)   Length of Paragraph;

    (g)   Ending of Paragraph.

Some examples of paragraphs are given below :

Example 1 : Parents are most competent to select their children's career.

It was a generally accepted principle until recently that the teacher is the most competent person to select a career of a child. This was, for the reason that he knew the student's social qualities and capabilities because of his daily contact with the student for many hours. This was possible when the student-teacher relationship was closer. The number of students in a class being limited, the teacher could give personal attention to each student. He could take interest in every student. This is no longer possible because the strength of students in a class has gone very high and in some cases, over 75 to 80. Parents, therefore, are the most competent to choose their children's career. They know the temperament, ambitions, inclinations, ability of their children. They can also take into consideration their own financial positions.

Example 2 : A Hungry Man is an Angry Man

Now-a-days it is a familiar picture of a mob of workmen on strike gathering outside a factory or an office shouting different slogans and abusing a strike. If a strike continues for a long period, it leads to hunger in the families of the workers. A hungry man becomes desperate and violent leading to disturbances. The peace of mind of himself, his family members and of the society in which he lives, is disturbed. There is no cure for hunger. For procuring food the man may resort to any means like beg, borrow or steal. The more pitiable sight is that of the rural people who are affected by famine. They are bodily, mentally and spiritually broken. At times ultimately lose even the strength to protest.

Example 3 : A paragraph on the various activities of a mine (Technical)

A mine usually works in three shifts. A miner comes to the mine in his normal clothes, puts them in the locker and changes into working clothes. He then collects a lamp and comes to the pit head. He is then checked to see that he does not carry with him any explosive or inflammable material. The miner enters the cage which is raised or lowered in the shaft. The shaft is controlled from the winding room. It is lined with hard material like brick, stone or concrete so that it may not crumble or crack.

When the cage reaches the bottom of the mine, the miner comes out. Here on all the sides there are tunnels and their roofs are supported by girders or propellors. Whenever possible, electric light is provided.

Example 4 : Television does more good than harm (Technical)

To consider negative results, we can say that there are certain advantages of having a television set in the house. Children and students in the house lose their concentration on studies and neglect the activities outside home i.e. sports, exercise etc. Elderly people become less social because they neglect their friends and relatives. They find it more comfortable to stay at home to watch television rather than visiting their acquaintances. Sometimes, undesirable films are shown which corrupt and pollute young minds.

Advantages of having a T.V. Set, outweights the disadvantages. General knowledge through visual effect becomes realistic. Inventions of science can be more clearly understood. The international events can be viewed with a special interest. People are acquainted with customs and traditions of countrymen helping to create international understanding. Television is welcomed all over the world.

Example 5 : Sunbathing

Sunbathing is recommended by physicians as an excellent way to improve one's physical conditions. When wisely engaged in, it builds up resistance to infections and tones up the whole system. Too much exposure may be injurious to health. Calcium in the bone structure may be depleted. A painful overdose of sunburn may develop into something really serious. Other dangers lurk in the blazing rays of the mid summer sun. Caution should be exercised when one is taking a sun bath, specially if he is not accustomed to it.

(2) COMPREHENSION :
UNSEEN PASSAGE FOR COMPREHENSION

COMPREHENSION

The main object of setting an unseen popular, science or technical passage is to assess the students' ability to grasp the meaning of the given passage and answer certain questions based on it. It, therefore, involves careful reading of the passage in order to answer the questions in exact and precise sentences. A single reading of the passage is not enough to understand the clear intention of the writer, theme or subject of the passage, attitude or approach of the author to the theme and finally the details and style of the passage. It has to be read again and again. If the passage is read several times (at least three times) it ensures better understanding of minute technical and scientific details of the passage with every additional reading.

In the first reading, the reader will be able to understand the broad view about the theme of the passage. The subsequent readings will help the reader to grasp significant points and minute technical and scientific details. This will facilitate the reader to answer any question set on the given passage. In short, good answers to the questions on the comprehension passage can be given only when :

(1) The reader fully understands the meaning of the passage;

(2) He should be able to express the meaning of the passage in his own simple and clear words; and

(3) He should properly understand the questions asked.

1. How to proceed for the preparation for Comprehension Test ?

The following points should be followed in order to do well in the comprehension test :

(a) Reading of the passage carefully for at least three times;

(b) The first reading should be rapid for understanding broad ideas;

(c) The second reading should be little slow keeping in mind two definite ends viz.

    (i) To confirm whether the impression after the first reading was correct and exact; and

    (ii) To underline the key ideas.

(d) The third reading should aim at understanding :

    (i) The subject of the passage; and

    (ii) What is said about the subject.

(e) The reader then should ask himself the following questions :

    (i) What is the main theme of the passage ?

    (ii) What does the author say about the subject ?

    (iii) Can I put in my own words what the author has to say ?

(f) Read the given questions carefully in order to classify them into :

    (i) Broad questions, and

    (ii) Particular and short questions.

(g) Link up each of the questions with the relevant portion in the passage.

(h) Prepare answers to the questions in your own simple and clear words. Do not attempt to copy out the words in the passage.

(i) Give attention to the grammar, punctuation and correctness of the language used in the answer.

(j) The answer should be brief, to the point and free from irrelevant details.

(k) Read the answers again to ensure
  (i) That they are clear, complete and to the point;
  (ii) If they need revision;
  (iii) If they need expansion or reduction;
  (iv) That there are no grammatical or spelling mistakes;
  (v) That the language is simple and direct.

(*l*) If there is a question on explaining the meaning of a word or phrase, the answer should be given in the same grammatical form.

(m) Give title or heading to the passage if it is asked. While doing so, the following hints may be kept in view :
  (i) Main character or object explained in the passage.
  (ii) Some saying or proverb illustrated in the passage.
  (iii) Heading reflecting the idea of the whole passage.

(n) Finally, ensure that all the questions given under the passage are answered and the answers are properly numbered.

A few model comprehension passages are given below. It would be useful if students attempt to answer the questions themselves and then compare them with the model answers given under these passages.

Example 1

Read the following passage and answer the questions given below :

We want to make you see bridges as we see them, not as mere prosaic objects of utility and economy, but as something far more significant and inspiring for a bridge is more than a thing of steel and stone. It is the embodiment of the efforts of human heads, hearts and hands. A bridge is more than a sum of stresses and strains; it is an expression of man's creative urge, a challenge and an opportunity to create beautiful structures. A bridge is the fulfilment of human dreams and hopes and aspirations. A bridge is the symbol of humanity's heroic struggle towards mastery of the forces of nature. A bridge is a monument of mankind's indomitable will to achieve.

Bridges symbolise the ideals and aspirations of humanity. They span the barriers that divide and they bring people, communities and nations into closer unity. They shorten the distances, speed up transportation and facilitate commerce. They carry their burdens so that the task of men may be lightened. They serve the needs of the lowest as of the highest. They are cooperative efforts of planners and workers, of science and skill. They embody the initiative and vision of communities in useful monuments dedicated to the welfare of future generations. They are vital links in the highway leading to the universal brother-hood of mankind.

Q.1 What do bridges symbolize ?

Ans. Bridges are the symbols of the efforts of human heads, hearts and hands. They express man's creative urge. They fulfil human dreams, hopes and aspirations. Bridges are also the symbols of humanity's heroic struggle towards controlling the forces of nature and mankind's will which cannot be surpassed.

Q.2 How do bridges lighten the task of men ?

Ans. Bridges take on the burdens which men would have been required to carry. Thus, they lighten the task of man.

Q.3   **What is a bridge an outcome of ?**

Ans.   It is an outcome of the co-operative efforts of a planner and workers, of science and skill.

Q.4   **Give the meaning of the following words :**

Ans.   Embodiment – symbol in concrete form.

Indomitable – which cannot be suppressed.

Q.5   **Use the following in sentences of our own :**

Ans.   **(a)**   **To serve the need:** Roads serve the need of mankind.

    (b)   **Creative urge :** Tall buildings are the symbols of man's creative urge.

Example 2 :

(a) **Read the following passage carefully and answer the questions given below in short and in your own words :**

An alloy steel is one in which other metals have been added to plain carbon steel to give it certain qualities for special uses. The addition of alloy metals increases the toughness and durability of steel and often makes possible a reduction of the weight of the steel product. There are three principal purposes for making steel alloy. They are, to correct or prevent defects in the final product, to impart some distinctive property to the steel and to form alloys for experiment and investigation. Some of the alloying metals are manganese, nickel, chromium, molybdenum, vanadium, tungsten and various combinations of these elements. The alloys give the steel properties such as hardness and resistance to heat required in tools and machinery.

Complex alloy steel contains more than two alloying metals, in addition carbon. A well-known variety is stainless steel, which contains chromium and nickel. Stainless steels do not rust, they withstand higher temperatures and resist most of the acids. They are used in the manufacture of cutlery and surgical instruments, and also for the kettles and piping used in the food industries.

Q.1   **What are the properties of stainless steel ? Where is it used ?**

Ans.   The most important property of stainless steel is that it does not rust and it can withstand high temperatures. In addition to these qualities, it can resist most of the acids. It is used for manufacturing pipes, kettles and is also used in food processing units.

Q.2   **What is alloy steel ?**

Ans.   An alloy is a substance obtained after the combination of one or more metals with another. In the case of alloy steel, other metals are added to plain carbon steel.

Q.3   **Name the different alloying metals. What properties do they give to   steel ?**

Ans.   Important alloying metals are manganese, nickel, chromium, molybdenum, vanadium, tungsten and various combinations of these elements. They give steel the desired properties like hardness and resistance against heat.

Q.4   **Why is steel alloyed ?**

Ans.   Steel is alloyed with other metals for three principal purposes viz. to correct or prevent defects in the final products to impart some distinctive property and to form alloys for investigations and experiments.

Q.5   **Form sentences of your own using the following :**

Ans.   **(a)**   **To Impart :** He is happy to impart training to his assistants.

    (b)   **In addition to :** In addition to his normal duties, he also co-ordinates the activities of the various departments.

Example 3 :

Read the following passage carefully and answer the questions given below :

The invention of the wheel marked an epoch of revolution in the fields of transport, industry, trade and commerce. All the means of transport in the present age such as railways, buses, motor-cars, tractors and bicycle are gifts of the invention of the wheel. The power generation machines, textile mills machines, printing press machines, flour rice and saw mills elevators and a large number of other machines that produced many essential commodities of daily use are fundamentally based on wheel power. Without this invention, our civilization would have remained just a primitive one. Imagine what our life would have been without these gifts of the wheel ! Our movements would have been almost totally restricted. The known world would have grown too large in the context of distance. Communities living in the various parts of the world would have remained isolated from each other. The exchange of goods and services all over the world would have been an impossibility.

If the inventions based on the wheels are biliterated, our life would relapse to that of a frog in the well. It would bring back the primitive agricultural age. The problems of time and distance would stare us in the face. The modern industries alongwith their large scale production of the beautiful fabrics, books and magazines and a large number of other things would have remained only a dream.

Q.1   **What are the gifts of the invention of the wheel ?**

Ans.   All sorts of means of transport for example buses, railways etc. are the gifts of the invention of the wheel.

Q.2   **What are the various things based on the wheel power ?**

Ans.   Many essential things of daily use like flour, rice and saw mills, elevators are mainly based on wheel power.

Q.3   **How would communities and exchange of goods etc. be affected without the wheel ?**

Ans.   Without the wheel, our movement as well as exchange of goods would be limited and there would be no progress.

Q.4   **What would not have come in reality in the absence of the inventions based on wheel ?**

Ans.   In the absence of wheel, our life would have remained just like that of uncivilized man. All the astonishing inventions are due to wheel; for example : large scale production of beautiful fabrics, books, magazines and a large number of other things.

Example 4 :

Read the following passage carefully and answer the questions :

Machines were made to be man's servants, yet he has grown so dependent on them that they are in a fair way to become his masters. And machines are very stern masters. They must be fed with coal and given petrol to drink and oil to wash with, and they must be kept at the right temperature. If they do not get their meals when they expect, they grow sulky and refuse to work or burst with rage and blow up and spread ruin and destruction all around them. So we have to wait upon them attentively, and do all that we can to keep them in good temper. Already we find it difficult either to work or play without machines, and a time may come when they will rule us altogether, just as we rule the animals. Machines save time and energy for us but we use them for making more and better machines. We must remember that machines themselves or the power that machines have given us are not civilization, but aids to civilization. Real civilization will come only when we learn to use these machines as instruments in the service of humanity at large and prevent them from being a means of luxury and power for a few.

Q.1	**How are we required to look after machines ?**
Ans.	We are required to look after the machines by feeding them with coal, giving oil to wash and petrol to drink. They must also be kept at the right temperature.
Q.2	**What happens when machines are neglected ?**
Ans.	Machines grow sulky and refuse to work when they are neglected.
Q.3.	**When do machines become a curse to mankind ?**
Ans.	When we find it difficult to work with machines they become a curse to mankind and ultimately they rule us.
Q.4	**When can we have real a civilisation ?**
Ans.	We can have real a civilisation when we can use these machines as instruments in the service of humanity.

Example 5 :

Read the following passage carefully and answer the questions :

The definition of progress shall always continue to be a debatable point. Certain evils do follow every technological advance. According to some, they overweight the good brought about by technological advance. Conservatives never appreciate the modern inventions and changes in our day-to-day life. They consider them as changes for the worse; others ignore the latent dangers and praise the advances. In the beginning, people were thrilled with the invention of the motor car. They enjoyed the facility of fast movement, visiting places, better roads etc. But as cars became more and more in number, common people faced the problems of traffic jam and parking, exhaust gas pollution, road accidents etc. For every life saved by a speeding ambulance, another is taken by a skidding car.

The television, though a very popular means of entertainment and information, it has killed the conservation and normal two-way contact between individuals.

Lately, development of pesticides and weed killers have led to larger production of food but they have tampered with the balance of nature. They have caused death and the sterility of innumerable birds and insects. D.D.T. is capable of causing irreparable damage to the tissues of human body.

The truth is that in regard to all advances we have failed to give sufficient thought to their long term results and consequences. We have failed to make proper analysis of immediate gains and long term harms affecting even the very existence of life in the world. In most cases, we are so excited by a new invention that we do not care to assess, even roughly, the difference between happiness and misery it will cause.

Q.1	**What is the difference between the views of conservatives and other people, regarding technological advances ?**
Ans.	Conservatives do not like the modern inventions and any change in daily routine due to technological advancement; whereas other people appreciate them.
Q.2	**How can cars save life and also take life ?**
Ans.	In case of an ambulance, it can take a serious patient to a hospital very speedily. But the same car takes human life in road accidents.
Q.3	**What is the disadvantage of television ?**
Ans.	The demerit of television is that it has reduced the conversation and social contacts among people.
Q.4	**What do we forget in our excitement about a new invention ?**
Ans.	In our excitement about a new invention is that we do not care to find out the difference between happiness and the misery it may cause.

> **Q.5   Give the meaning of :**
> **Ans.   (a)   Debatable point :** The point about which people carry different views.
>   **(b)   Irreparable damage :** A damage which cannot be made good by any means.

Exercise 1 :

As you know, majority of the people in India cannot read or write. They are illiterate. This is not their fault. They have never had the chance to learn and read. But we know, too, that our country cannot progress as it should, if the majority of the people are ignorant and uneducated. If we are to be the best kind of citizens we must be educated. We must atleast be able to read books and newspapers and magazines. Now those of us, who have had the chance to go to school and to be educated, have been given something which many of our neighbours have been deprived. They, therefore, need our help. In this matter we are in position to help them, if we are willing to do so. We should show our neighbourliness if we found our next door neighbour lying at the side of the road unable to move because he had broken his leg. Similarly, we should show our neighbourliness when we find him unable to make progress because he cannot read.

Read the above passage carefully, and answer the following questions :
1.   Why is our country not making progress as it should ?
2.   How can we remove illiteracy from our country ?
3.   What is your duty to your illiterate neighbour ?
4.   What should you do if you found a person lying injured on the road ?

Exercise 2 :

Of man's earliest inventions, we know very little. The first may have been the use of stone to crack a nut. The next was possibly the use of stock to strike an enemy. Once man found that stock and stone were useful, it was only a step further to the making of a rude weapon by fastening a stone to the end of a stock.

Man used sticks and stones long before he dared to meddle with fire; for the early men, like all wild creatures, dreaded fire. Fire, of course, existed for lightning must sometimes have set the forests ablaze just as it does today; and in those days volcanoes were much more active than they are now. The forgotten hero who first dared to tame fire to his own use was the greatest of early inventors; for once man had fire he was master of all lower creatures.

Questions :
1.   What would have been man's earliest inventions ?
2.   Why did man not tame fire as early as he came to know the use of sticks and stones ?
3.   What is there to show that fire existed long, long ago ?
4.   Why does the author of this passage call the first tamer of fire the greatest of early inventors ?

Exercise 3 :

The basic theory according to which the hydro-electric plant operates is simple. The higher the elevation from which a weight is allowed to fall the faster it will be moving when it reaches the bottom. If a column of water, falling free through the air, is allowed to drop on the vanes of a waterwheel it will, of course turn the wheel round. The old waterwheels are operated as simply as this. The higher the fall on the wheel, the more power it was able to deliver. The pressure would vary as the head varied.

After having been almost abandoned as obsolete and supplemented by steam engines, the old water-wheel has reappeared in a new and superior form under the name of water turbine. Connected with the wheel is a *generator* of electricity, popularly called dynamo. The *turbine* is a wheel which combines the use of impact of a falling mass of water with the pressure caused by the height of main body of water. It makes use of the pressure by means of its curved vanes upon all of which the water acts simultaneously.

The difference of elevation of the surface of the main body and that of the water which leaves the turbine at the *tailrace,* is the total head. As some of the pressure is lost because of friction in pipes and other passages (friction head) what is left to do the actual work is called *effective head*.

Read the above passage carefully, and answer the following questions :
1. What is water turbine ?
2. On what principle does the water-turbine work ?
3. What is the use of water turbine ?
4. What is meant by 'effective head' ?
5. What is the basic theory on which the hydro-electric plant operates ?
6. What is the new and superior form of water wheel ?

Exercise 4 :

Metallurgy is the most important science dealing with the separation of metals from their ores. It is not of recent origin for it was well-known to the ancients; modern men have only improved upon ancient formulas and methods.

Metals are found in three classes of ores; those in which the pure metal occurs either in grains or loose nuggets; those in which the metals occur as oxides; and those in which the metals occur as sulphides. Ores of the first class need little treatment. This consists in crushing the rock and separating the loose metal from it. The metal is then united into larger masses by smelting. The oxides constitute by far the largest class of ores and it is from these that the supply of iron, lead, tin, copper and zinc is mostly obtained. Most of these ores can be reduced by smelting with a flux, as in the case of the manufacture of pig iron. The sulphides are more difficult to treat, and some of them require several processes before the metal is obtained.

In general, the treatment of this class of ores is as follows : The ore is crushed, and the metal bearing portion is separated by running the crushed ore over vibrating tables, over which water is running. The particles containing the metal, being heavier than the others, settle at the bottom and form what is known as the *concentrate*. This concentrate is dried and roasted to drive off the sulphur. The ore is then smelted; it yields an impure metal, which is purified by repeated meltings. Copper ores containing sulphur are reduced in this manner.

Read the above passage carefully and answer the following questions :
1. What is metallurgy ?
2. What are the three classes of ore ?
3. Which is the largest class of ores ?
4. How are the sulphides treated ?
5. What is concentrate ?
6. How do we get purified metal ?

Exercise 5 :

Dry cells are a costly way of producing current, and when powerful currents are needed for long periods, a different kind of cell is used. This can be 'recharged' with electricity when its supply is exhausted, by being joined for a few hours to the electric supply from a power station. The cells are called storage cells, but what they store is not electrical energy, but chemical energy. An ordinary storage cell is a glass or plastic box containing a mixture of sulphuric acid and water, with one 'positive' and two 'negative' plates dipping into the acid. The plates are hollow lead boxes pierced with holes to allow the acid to enter freely inside. The positive plate is filled with a brown paste of lead peroxide and sulphuric acid. The negative plate is filled with spongy lead. When the cell supplies electrical energy, the lead peroxide in the positive plates reacts with the acid to form lead sulphate, and so does the lead of the negative plate, the reverse changes take place when the cell is charged. Small batteries each having six storage cells are used in cars to work the headlights, the starter motor, the traffic signals and so on.

Electrical energy is also converted into chemical energy in many factory processes when an electric current is passed through liquids to produce the chemical action called electrolysis. Electrolysis is used, for example, to electroplate spoons with silver or car radiators with chromium and to purify copper, gold, silver, zinc and other metals. Electrolysis is also used to extract the metals like aluminium, magnesium, calcium, sodium and potassium from their ores.

Read the above passage carefully, and answer the following questions :
1. What is meant by recharging ?
2. What does an ordinary storage cell contain ?
3. Where is the small battery cell used ?
4. How is the electric energy converted into chemical energy ?
5. What is electroplating ?
6. What are the uses of electrolysis ?

Exercise 6 :

Under an old apple tree in my garden, a strange thing, has happened. A tiny red mite, dropping through the air from the leaves above, has landed on the open page of a note book. It runs over the paper in curves. Each time it comes to the wet ink of a freshly written lines it pauses for a long time as though on the bank of an impassable river. Then, I notice its body is turning darker, is losing its brilliant red. The mite is drinking the carbon ink ! Eight times it stops and eight times it drinks from the black stream of the ink. Only its legs retain their original colour. Its round body is now swollen with carbon ink. Yet the black mite is as lively as the red mite had been. The ink in its system had neither slowed it down nor upset its digestion.

Mites imbibing ink, cockroaches consuming glue, termites eating wood – are only a few of many instances of small creatures that live on what would prove a deadly diet for larger animals. The strength of these insects lies in the bewildering variety of their diet. Because their feeding habits are so varied, because the amounts of food they require are so small, such creatures are able to endure privations that woull kill far stronger animals.

We can crush a butterfly or a beetle between a thumb. We are giants in comparison. But the smaller creatures of the earth possess other special kinds of strength. They survive extreme and sudden changes in temperature. They retain life inspite of great injuries. They reproduce with a fertility unequalled among larger animals. They are uninjured by long falls; and they possess the strength of senses that are abnormally keen. Honeybee can see ultraviolet light. The bedbug has such an amazing sense of temperature that it can detect a difference as slight as one degree centigrade.

Questions :
1. What did the mite do when it came to the wet ink on a freshly written line ?
2. What change took place in the mite after it had drunk the ink ?
3. Write four arguments that the writer gives to prove that though we are giants in comparison, the smaller insects have special kinds of strengths.
4. Give a suitable title to the text.
5. Find single words in the passage which have the following meanings :
   (a) entail, (b) devour.

Exercise 7 :

Our country gave birth to a mighty soul and he shone like a beacon not only for India but also for the whole world. And yet he was done to death by one of our own brothers and compatriots. How did this happen ? You might think that it was an act of madness, but that does not explain this tragedy. It could only occur because the seed for it was sown in the poison of hatred and enmity that spread throughout the country and affected so many of our people. Out of that seed grew this poisonous plant. It is the duty of all of us to fight this poison of hatred and ill-will. If we have learnt anything from Gandhiji, we must bear no ill-will or enmity towards any person. The individual is not our enemy; it is the poison within him that we should fight and which we must put an end to.

Questions :

1. Who is 'the mighty soul' referred to in this passage ?

2. Why was he done to death ?

3. What do we learn from Gandhiji ?

4. What should we fight against in the individual ?

5. What is the poison referred to in this passage ?

6. Write a suitable title for the passage.

7. Write the words from the passage which mean the following (any two) :

    (a) Unhappy event or end

    (b) Powerful

    (c) Signal fire or torch on a hill.

Exercise 8 :

Here is a scientific experiment on the homing of birds, the facts of which are quite certain. A few years ago seven swallows were caught near their nests at Bremen in Germany. They were marked with a red dye on some of their white feathers, so that they could be easily identified. Then they were taken to Croydon by plane – a distance of about 700 miles and set free. Five of them flew back to their nest at Bremen. How did the birds find their way on that long journey, which they had never made before ? That is a great puzzle. It is no good saying that the swallows (or dogs) have a sense of direction or an instinct to go home. These are just assumptions and explain nothing. We want to know exactly what senses the animals use to find their way, how they know in which direction to go until they see familiar landmarks. Unfortuantely, practically no scientific experiments have been made on this question.

Perhaps migrating birds are the greatest mystery of all. Swallows leave England in August and September, and they fly to Africa, where they stay during the English winter. The swallows return to England in the spring to nest. A lot has been found out about the journeys of migrating birds by marking the birds with aluminium rings put on one leg with an address and a number put on the ring.

It has been suggested that birds can sense the magnetic lines of force stretching from the north to the south poles of the earth, and so direct themselves. But all experiments hitherto made to see whether magnetism has any effect whatsoever on animals, have given negative results. Still, where there is such a biological mystery as migration, even improbable experiments are worth trying.

Questions :

(a)   Why is the migration of birds such a mystery ?

(b)   How are the birds marked for experimentation purposes ?

(c)   What has been suggested about the homing birds ?

(d)   Make sentences with the following phrases so as to bring out their meanings.

    (i)   sense of direction   (ii)   scientific experiments

    (iii)   magnetic lines   (iv)   biological mystery

Exercise 9 :

Despite its extraordinary name Ergonomics is not just something which has been dreamed up by experts to confuse any business. It is a practical science which can make and is making a valuable contribution to industry. Research being carried out in this field can increase efficiency, improve product design, and make employees more contented.

Broadly, Ergonomics is the study of the relationship between man and his working environment. It involves designing or redesigning machines and equipments so that due regard is given to the capabilities and limitations - physical and psychological of the human beings who have to use or operate them. In contrast to the more common, traditional policy of 'fitting the man to the machine' by selecting and training the best man for each particular job, the ergonomist endeavours to fit the machine to man - any man. His object is to enable a person of ordinary abilities to carry out his tasks safely and efficiently with a minimum of instructions.

The ergonomist himself needs knowledge of a number of sciences and techniques in order to carry out his work. These include Psychology, Physiology and Anatomy. In addition, of course, he must know something about engineering and work study techniques.

Although the word 'ergonomics' was born as recently as in 1949, and the concentrated study of this new science dates back to scarcely two decades, its origin can really be attributed to early nineteenth century astronomers. They started the ball rolling when they discovered, that, contrary to general belief, human reactions were not instantaneous. Observers engaged in measuring the speed of star movement across the lenses of their telescopes did not agree in the times they reported. So a new factor had to be taken into consideration, the time taken by human mind to react to the message from the eyes.

Questions :

(1)   What is ergonomics ?

(2)   How does it help business and industry ?

(3)   What exactly does ergonomics involve ?

(4)   What was the traditional policy before ergonomics came on the scene ?

(5)   What is the job of an ergonomist ?

(6)   What are the other sciences which the ergonomist should know?

(7)   Trace the origin of ergonomics.

(8)   Give a suitable title.

QUESTIONS

Write paragraphs on the following subjects.

(a) Game

(b) Football

(c) Railway Station

(d) My Institute

(e) A Holiday

(f) Examination

(g) Theatre

(h) Zoological Park

(i) Library

(j) Workshop

(k) Tool box

(l) Bus Stop

(m) Work-shop

(n) I.C. Engine.

Part – IV

VOCABULARY BUILDING

(1) Word Formation
(2) Technical Jargon
(3) Use of –
    (A) Synonyms
    (B) Antonyms
    (C) Homophones / Homonyms
    (D) Same Words used as Different Parts of Speech
(4) Use of Contextual Words in a given paragraph.

[Marks 12]

VOCABULARY

The word vocabulary is broadly defined as a person's knowledge about words. For acquiring mastery of any language it is quite essential to increase one's word-power. In other words, more the number of words known to a person, more rich is his vocabulary. For enriching one's vocabulary there are numerous ways one can adopt. The first and foremost is to remember as many words as possible and make their frequent use in speaking and writing which helps one to remember those words with little or no effort. Once a person gets accustomed to this practice, he can shift the emphasis to other aspects of increasing word power.

The further step in this direction is to get acquainted with different derivatives. By definition, derivatives are the words which are not original but obtained from other words by addition of affixes which are the meaningful elements but cannot be used independently. Affixes are either in the form of prefixes like pre -, equi -, in - or suffixes like ment, - ish, - ly etc. For example –

pre + planned	=	preplanned
in + sufficient	=	insufficient
equi + lateral	=	equilateral
establish + - ment	=	establishment
blue + - ish	=	bluish
Friend + - ly	=	friendly

Sometimes, a word is moved from one grammatical class to another without adding a prefix or suffix. Such derivatives are called 'zero derivatives'. Derivatives are formed not only from simple words but also from derivatives. English language is so typical that there are words which carry different meanings or have different pronunciations eventhough their spelling is the same. Such words are called as 'homographs. There are also words which are spelt or pronounced in the

(4.1)

same way but carry different meaning. For example, the word 'see' is spelt and pronounced in the same way. But when used as a verb it means 'to perceive or become aware of something by using the eyes and if used as a noun it means office or jurisdiction of a bishop or archbishop.

Vocabulary can also be enriched by studying different words carrying the same meaning i.e. synonyms, words with opposite meaning i.e. antonym compound words, words that can replace a group of words, phrases, idioms, proverbs etc. A deliberate attempt towards the above aspects of words would certainly contribute to increase one's treasure of word-power. We will study some of the above features one by one.

(1) WORD FORMATION

In English language, word formation is concerned with derivatives, formed from simple words (roots) by adding either prefix or suffix or both. Simple words which are also called as **Primary Words** belong to the original words in the language. They are not derived or compounded or developed from other words. Derivatives are either **Primary Derivatives** or **Secondary Derivatives**. Primary derivatives are formed by making change in the body of the simple words. It should be noted that, the Past Tenses of primary verbs formed by internal change do not fall in the category of primary derivatives. Some of the primary derivatives are listed below.

(I) PRIMARY DERIVATIVES

(a) Nouns from verbs and Adjectives

Verb	Noun	Verb	Noun
Advise	Advice	Gape	Gap
Bind	Bound	Grieve	Grief
Bless	Bliss	Live	Life
Break	Breach	Lose	Loss
Choose	Choice	Prove	Proof
Deem	Doom	Sing	Song
Float	Fleet	Speak	Speech
Wake	Watch	Strike	Stroke

Adjectives	Nouns
Dull	Dullness
Hot	Heat
Proud	Pride

(b) Adjectives from Verbs and Nouns

Verb	Adjectives	Verb	Adjectives
Float	Fleet	Milk	Milch
Lie	Low	Wit	Wise

(c) Verbs from Nouns and Adjectives

Nouns	Verbs	Nouns	Verbs
Bath	Bathe	Glass	Glaze
Blood	Bleed	Grass	Graze
Brood	Breed	Knot	Knit
Cloth	Clothe	Price	Prize
Drop	Drip	Tale	Tell

(ii) Secondary Derivatives

As stated earlier, the secondary derivatives are obtained by using affixes. Affixes may be prefixes or suffixes. An addition to a word in the beginning is a prefix while to the end it is a suffix. Some of the prefixes and suffixes used for getting secondary derivatives indicating their purport and words obtained by adding them are given below -

(a) Prefixes (Inclusive of Latin and Greek Prefixes) in common use

Prefix	Purport (to mean)	Words obtained by adding the Prefixes
A	on, in	aboard, ashore, asleep
A	Out, from	arise, awake, alight
A (an)	Without, not	athiest, apathy, anarchy
Ab (a, abs)	From away	avert, abstract
Ac (ad, af, ag, al, an, ap, ar, as, at, a)	to	accord, adapt, affect, aggrieve, allege, announce, appoint, arrest, attach, avail
Ambi (amb, am)	on both sides	ambiguous, ambivalent, ampute
Amphi	around, on both sides	amphibious, amphitheater
Ana	up, back	anachronism, analysis
Ante (anti, an)	before	antedate, anticipate
Anti (ant)	against	antipathy, antagonist
Arch (archi)	chief	archduke, archbishop, architect
Auto	self	automobile, autonomy, autograph
Be	by	beside, betimes
Bene	well	benefactor, benefit, benevolent
Bis (bi, bin)	twice, two	bisector, binocular, bigamy
Circum (circu)	around, round	circumference, circumlocutory, circular
Con (cog com, cor)	together, with	Consent, colleague, combine, correlate
Contra (counter)	against	contraception, counterfeit, counteract
De	down	devaluation, degenerate, deplete
Demi	half	demigod
Di	two, double	dioxide, dilemma
Dia	through	diameter, diagonal
Dis (diff, di)	apart, reverse, opposite, negative	dishonest, disfigure, differ, divide

En (em)	in	enclosure, emblem, encyclopedia
Epi	upon	epitaph, epilogue
Eu	well	eulogy, euphony, eugenics
Ex (ec, ef, e)	out of	extract, exodus, eccentric, effect, educe, extract
For	thoroughly	forgive, forget, forbear
Fore	before	foregone, forefather, forecast
Hemi	half	hemisphere
Homo (hom)	like	homogeneous, homosexual, homophone
Hyper	over, beyond	hyperbole, hypertension, hypercritical
Hypo	under	hypocrite, hypothesis
In (im, il. ir, en, em)	in into	income, inland, illustrate, immerse, irrigate, enact, embrace
In (il, im, ir)	not	indecent, illegitimate, immature, irreligious
Male (mal)	ill, badly	malpractices, malnutrition, malevolent
Meta	of change	metaphysics, metabolism, metamorphosis
Mis	wrong, wrongly	misguide, misuse, misfortune, misdeed
Mono	single, one	monologue, monogamy, monorail
Non	not	nonentity, non-fiction, nonsense
Over	above, beyond	overact, overdose, overcharge
Para	by the side, beside	parasite, paramilitary, parameter
Peri	around	periphery, perimeter, periscope
Philo (phil)	love, liking, fond of	philology, philosophy, philanthropy
Pre	before	prefix, pre-historic, prenatal
Pro	for, in favour of, supporting	prophesy, pro-German, programme, pro-chancellor
Quasi	to a certain extent, not really	quasi-permanent, quasi-rent
Re	back, again	regain, rewind, reclaim, refund, rewrite, regenerate
Retro	backward	retrospective, retrograde
Semi	half	semi-skilled, semi-precious
Sub (suc, suf, sug sum, sup, sur, sus)	under	subsidy, succeed, suffer, suggest, summon, support, surmount, sustain
Super	above	supernatural, superfluous, superintendent, superfast
Syn (sym, syl, sy)	with, together	synonym, sympathy, syllable
Trans (tra, tres)	across	transworld, traverse, trespass
Un	not	unkind, unbelievable, unruly
Un	to reverse an action	unfold, undo, unveil, untie
Under	beneath, below	underground, undergo
Vice	in place of	viceroy, vice captain, vice-president

(b) Suffixes (inclusive of Latin and Greek) in common use

Suffix	Purport	Words with these suffixes
Of Nouns	**Denoting agent or doer of a thing**	
- ain (-an, -en, on)		Chieftain, artisan, citizen, surgeon
- ar (-er, -eer, ier, - ary)		scholar, preacher, engineer, financier, missionary
- ate (-ee, -ey, -y)		advocate, trustee, attorney, deputy
- er (-ar, -or, -yer)		painter, driver, baker, sailor, lawyer
- or (-our, -eur, -er)		emperor, saviour, amateur, interpreter
-ster		spinster, songster, punster
- ter (- ther)		daughter, father
- age		bondage, marriage, leakage
- ance ence)		assistance, brilliance, excellence, intelligence
- cy		intimacy, leniency, fancy
-dom		stardom, kingdom, wisdom
- hood(- head)		brotherhood, manhood, godhead
- ice - ise)		service, cowardice, exercise
- ion		opinion, selection, union
- lock (- ledge)		knowledge, wedlock
- ment		treatment, judgement, punishment
- mony		testimony, matrimony, parsimony
- ness		fitness, business, brightness
- red		kindred, hatred
- ship		kinship, guardianship, lordship
- th		health, wealth, growth
- tude		multitude, magnitude, fortitude
- ty		frailty, cruelty, credulity
- ure		pleasure, forfeiture, verdure
- y		treasury, victory, misery

	Forming diminutives	
- cule (-ule, - sel, - cel, - el, - le.)		molecule, globule, damsel, parcel, chapel, circle
- el (- le)		kernel, satchel, girdle, handle
- en		chicken, maiden, kitten
- et		lancet, trumpet, owlet
- ette		cigarette, coquette
- ie		dearie, birdie, lassie
- kin		lambkin, napkin, bumpkin
- let		booklet, bracelet, leaflet
- ling		darling, duckling, weakling
- ock		hillock, bullock
- ary (-ery, -ry)		library, treasury, nunnery, dairy, dispensary
- ter (-tre)		theatre, cloister
Of Adjectives		
- al	belonging to	formal, legal, international, global
- an (- ane)		human, mundane, humane
- ar		similar, regular, familiar
- ary		honorary, contrary, temporary, necessary, customary
- ate		temperate, fortunate, obstinate
- ble	having the quality	capable, culpable, sensible, laughable
- ed	having	gifted, talented, learned
- en	made of	wooden, golden, wollen, earthen
- esque	inthe style of, way of	picturesque, grotesque
- id	containing	humid, arid, lucid
- ile	like, belonging to	fragile, servile, juvenile
- ine	of, concerning	masculine, divine, canine
- ish	somewhat, like	reddish, childish, girlish, boorish
- ive	with the quality of	active, attentive, constructive
- lent	with the quality of	indolent, turbulent, virulent

- less	free from without	fearless, senseless, hopeless
- ly	like, having quality of	godly, humanly, manly
- ose(ous)	full of, possessing	virtuous, dangerous, copious, verbose
- some	with the quality of	gladsome, troublesome, wholesome, quarrelsome
- ward	inclining to	forward, backward, wayward
- y	with the quality of, possessing	healthy, wealthy, greedy, thirsty, dirty
Of Verbs		
- ate		assassinate, captivate, terminate
- en	causative, forming transitive verbs	weaken, sweeten, strengthen
- er	intensive or frequentative	glitter, chatter, flutter
- fy	to make	simplify, purify, qualify, terrify
- ish	to make	publish, furnish, punish, banish
Of Adverbs		
- ly	like	wisely, shrewdly, boldly
- long		headlong, sidelong
- ward (– wards)	turning	homeward, upward, downward
- way (- ways)	to towards	straightway, anyway, sideways
- wise	manner, mode	likewise, otherwise
Greek Suffixes		
- ic (- ique)	possessing the quality of concerned with, related with, dealing with	unique, angelic, cynic
- ist		artist, chemist
- isk	of	
- ism (- asm)		asterisk, obelisk patriotism, despotism,
- ize	possessing the quality of to make, to show	enthusiasm civilize, criticize, sympathize
- sis (- sy)		crisis, analysis, poesy, heresy
- e (- y)		catastrophe, monarchy, philosophy
	belonging to	

(iii) Compound Words

Compound words are those words which are formed by combining the different parts of speach viz. nouns, adjectives, adverbs, gerunds, prepositions. The compound words we get by these different combinations are mostly nouns, adjectives and verbs and are recognized as **Compound Nouns, Compound Adjectives** and **Compound Verbs**.

(a) Compound Nouns

Compound nouns are obtained by the combinations like

(i) Noun + Noun

Ex. armband, artwork, batsman, chairman, cowboy, doorbell, dockyard, dashboard, filmstar, godfather, gold-mine, hammerman, headgear, ink-pot, juke-box, keyboard, landlord, milkmaid, newspaper, nightdress, postcard, quartermaster, road roller, railway, screwdriver, silkworm, sound-box-, sunglasses, teapot, tablespoon.

(ii) Adjective + Noun

Ex. big-bank, broadsheet, black-board, deadlock, nobleman, sweetheart, fineart, halftruth, stronghold, highclass, lowland, quickstep, slowcoach, tenderfoot.

(iii) Verb + Noun

Ex. breakfast, cut-throat, cutpiece, daredevil, pickpocket, hangman, spendthrift, telltale.

(iv) Gerund + Noun

Ex. dancing-club, bathing-beauty, hiding-place, singing-chair, burning-train, looking-glass, walking-stick, booking-office, cooking-class, writing-table, scribbling-pad, dining-hall, stepping-stone, working-class, swimming-pool, smiling-face.

(v) Adverb (or Preposition) + Noun

Ex. afterthought, outswinger, inswinger, foresight, outlaw, bypass, off-beat, outside, overdose, underestimation, onlooker, inside.

(vi) Adverb + verb

Ex. backcomb, backwash, outcry, income, outcome, forebear, forecast, outfit, outflow, outlay, upkeep.

(vii) Verb + Adverb

Ex. spin-off, die-hard, drawback, lock-up, send-off, go-between.

(b) Compound Adjectives

Compound adjectives are obtained by the combination of –

(i) Noun + Adjective (or participle)

Ex. heart-warming, note-worthy, headstrong, homesick, lifelong, worldwide, snow-white, blood-red, pitch-dark, skin-deep, purse-proud.

(ii) **Adjective + Adjective**

Ex. red-hot, blue-black, dull-grey, white-hot.

(iii) **Adverb + Participle**

Ex. everlasting, never-ending, well-dressed, down-hearted, long-suffering, inborn, outspoken.

(c) Compound Verbs

Compound verbs can be obtained by the combination of -

(i) **Noun + Verb**

Ex. backbite, earmark, browbeat, waylay, typewrite.

(ii) **Adjective + Verb**

Ex. Safeguard, fulfill, whitewash.

(iii) **Adverb + Verb**

Ex. outcry, outwit, undertake, overthrow, upset, ill-use, ill-breed, in-breed, overtake, over-joy, over-act.

In most of the compound words, the first word tends to modify the second word. The accent is given upon the modifying word when the combination is complete. It may be noted that, when the two elements of the compound word are partially blended, a hyphen is put between them and the stress falls equally on both the words or elements.

(2) TECHNICAL JARGON

In the modern field of education, there are a number of technical branches which use their particular terminologies. Many words in these fields are such that they are not being understood by people at large. Real meaning and purport of these particular words in a particular terminology would be properly understood only if one acquires specialized knowledge of that particular science or art. The terminology used by different specialized fields can be termed as technical jargon. Although, people at large may not understand meanings of all the terms used in different fields of science or art, it becomes essential for the students of those particular fields to become acquainted with these technical jargons, without which their expressions in their subjects would not carry proper weightage. The language used without technical jargons in the fields of science or art may lead to verbosity.

USE OF POSITIVE TONE

Languages greatly differ from one another. Hence, it is often difficult for foreigners to use another language perfectly well. The mistakes are generally committed due to several peculiarities of that language. Mistakes may occur because of wrong choice of an article, wrong position of a word, wrong sequence of clauses, wrong sequence of tenses, wrong use of attitude etc. In the modern days of consumer's sovereignty, utmost care is required to be taken to please the customers by using polite and convincing language. Every firm or business organization has its clientale which it has to maintain by proper, polite and prompt service. No firm can afford to be rude or indifferent to its customers. It therefore, becomes obvious to use proper language while addressing individual customers or business firms. Proper etiquettes and manners are required to be followed in this respect. Particularly, the use of 'you attitude' becomes a must.

The important means of communication is language. It must therefore, be used properly. Words with different shades of meaning should be used very carefully. For example, the word 'record' can be used as a verb as well as a noun. Similar is the case with many words which are given in this chapter earlier. There are also words with difference in stress or pronounciation. Such words need to be used very carefully. The words like sight, site, or cite can cause misunderstanding while speaking. This has to be carefully avoided while using technical terms because such terms are not properly understood by people at large. Thus, every word, every term, every construction of sentences, choice of attitude occupies significant importance in business communication in particular. Some of the examples would make students understand the underlying principle using correct tone and language.

It is essential that business letters in particular should be written in the right tone. Imperative and harsh language should invariably be avoided to make correspondence pleasing, convincing and motivating. Business letter must be in warm and friendly style, positive in approach, concise in content and polite in language. Although the business letters hitherto were a little pedantic or indifferent in tone, the style these days is considerably changing. They are becoming more and more informal in order to attract and please customers. Business letters are now becoming worth reading from the point of view of propaganda and attractiveness. They are no more tedious and boring to read.

I.   Following the principle of 'You attitude'

Read the following sentences :

(i)   Unless the payment is received you cannot get the goods.

(ii)   You should take the delivery of your consignment immediately, otherwise you will forfeit your right.

(iii)   We have reminded you several times. But you do not reply.

(iv)   You should be alert when you receive our warning.

(v)   You should speak less and work more.

The above sentences sound little harsh and do not follow the general principle of politeness in writing business letters. These sentences can better be written as follows to sound more appropriate.

(i)   The goods will be delivered on receipt of payment please.

(ii)   Delay in taking delivery of the consignment would lead to forfeiture of your right over it.

(iii)   No reply has been received even after many reminders. Please attend to it.

(iv)   Be alert on receiving our warning.

(v)   Speak less, work more is the better policy.

II. Following the principles of Correspondence

The computer you have supplied has been a headache for me. The screen does not display good quality. The key-board is useless giving wrong signals Your after sale service is not proper. You require several reminders. Your representative utterly fails to carry out prompt repairs. I am repenting for purchasing this unit from you.

Solution :

The above complaint letter from a customer does not follow the principles of correspondence. This letter can be better written as follows :

'The computer you have supplied needs better service and attendance. It does not function properly and affects my work. It is better if you instruct your customer service department to be more prompt, active and well-versed in their task. Your trouble-shooting staff perhaps may have to attend refresher courses'.

Following are some more examples of this type :

1. We cannot entertain your complaint unless you personally come to our workshop.
2. No money will be refunded once the goods are sold. We may consider to replace them.
3. We always try to satisfy our customers. It is our main objective in carrying out our business.
4. We only accept cash. If you want to pay by cheque you will have to pay additional amount of Rs. 5 for each cheque towards commission charges.
5. We assure the delivery of goods at your doorstep at any locality without charging anything extra.

The tone of the above sentences can be made more positive and pleasing if written as follows :

1. It would be appreciated if you make a convenient visit to our workshop to sort out your grievance.
2. Goods will be replaced if defective. No cash refund please.
3. Customers' satisfaction is our sole motto.
4. Cash payments preferred. If payment is by cheque please add Rs. 5/- extra for each cheque, towards commission charges.
5. We assure free delivery of goods in any corner of the city.

III. Eliminating jargon and verbosity

Take the example of the following application for leave from an employee :

There are many private affairs, I am required to look into at the place of my birth where my parents are living. It will be very kind and obliging of you, if I am granted leave. I shall certainly and surely attend to the duties immediately after the work is done.

The application would sound better and appropriate if written like this –

A I am required to attend certain domestic work at my native place kindly grant me three days leave. I shall promptly resume duties after the stipulated period of three days.

Read the following examples also :

1. After the chief guest arrived he was given a warm welcome. Then he delivered his enlightening and scholarly speech to the students. The chief guest delivered his address after a warm welcome.

2. There were very hot discussions. Everyone tried to emphasise his own view point and dominate the meeting. Ultimately after a long time they finally decided to come to a conclusion and make a unanimous resolution. Thus they passed a resolution to end the prolonged meeting.

   A unanimous resolution was passed after hot discussions over the issue.

3. The team included such players who were good bowlers, good batsmen and good fielders. They were also physically quite good and capable. As a result the captain was very enthusiastic and expected to win the match. As the team consisted of physically fit all rounders, the captain was sure to win the match.

4. The sculptor began to admire the statute more and more and to imagine that it was indeed like one alive. He began to hang trinklets and jewels on the lovely statue and slip rings on her fingers and talk to her as if she were alive. Silks too he bought to clothe her limbs and flowers of many hues he laid at her feet.

   The sculptor adored the statue as if it was as living being and decorated it in all possible ways.

5. Look at the situation of women in our country. Almost 45 years after independence, we still have illiteracy, malnutrition, exploitation, discrimination, offences of rape, dowry deaths, violence in the home and outside. Repeated child births, overwork and neglect of their health care requirements put them among the highest mortality rates in the world. Even after 45 years of independence the conditions of Indian women is very precarious.

(3) USE OF

(A) SYNONYMS AND (B) ANTONYMS

As per the 'Oxford Advanced Learner's Dictionary', synonyms is ' a word (or phrase) with the same meaning as another in the same language, though perhaps with a different style, grammar or technical use '; and antonyms is a word that is opposite in meaning in the same language. Some of the commonly used words with their synonyms and antonyms are given below - Students are advised to carefully go through them and make self-practice to use them frequently in their own sentences. This will help them positively to increase their word power.

Word	Synonym	Antonym
Abnormal	Uncommon, Rare	Common, Normal
Absurd	Ridiculous, Foolish	Rational, Sensible
Abundant	Ample, Bountiful	Inadequate, scarce
Accept	Receive, Approve	Refuse, Reject
Accuse	Blame, Censure	Admire, Appreciate
Admire	Respect, Appreciate	Despise, Condemn
Advantage	Benefit, Boon	Disadvantage, harm
Agree	Accept, Consent	Disagree, Decline, Deny
Appreciate	Admire, Esteem	Despise, Depreciate
Attack	Assault, Charge, Encounter	Defend, Guard, Fortify
Backward	Retarded, Sluggish	Forward, Progressive
Bad	Evil, Corrupt, Wicked	Good, Chaste, Decent
Beautiful	Charming, Attractive	Ugly, Repulsive, Graceless
Benefit	Advantage, Boon	Loss, Damage
Best	Excellent, Highest	Worst, Meanest
Blunt	Dull, Flat, Rude	Sharp, Pointed, Polite
Borrow	Obtain, Receive, Take	Lend, Give, Deliver
Brave	Courageous, Gallant	Weak, Coward
Bright	Shining, Intelligent	Dull, Ignorant
Buy	Purchase, Procure	Sell, Dispose

Cheap	Worthless, Inferior, Low	Worthy, Superior, Costly
Cheerful	Delightful, Gay, Happy	Gloomy, Dull, Unhappy
Clever	Sharp, Talented, Skilful	Dull, Stupid, Fool
Competent	Able, Efficient, Skilful	Incapable, Incompetent
Condemn	Accuse, Blame, Censure	Appreciate, Praise
Constant	Firm, Perpetual, Stable	Flexible, Unsteady, Variable
Correct	Right, Accurate, True	Wrong, Incorrect, False
Courage	Boldness, Valour, Bravery	Cowardice, Fear, Timidity
Cruel	Brute, Callous, Unkind	Merciful, Kind, Humane
Curse	Bane, Malediction	Boon, Benediction
Damn	Condemn, Denounce, Curse	Admire, Praise, Bless
Dark	Black, Gloomy, Ebony	White, Bright, Clear
Decrease	Reduce, Lessen, Diminish	Increase, Augment, Enhance
Defeat	Failure, Retreat, Collapse	Victory, Triumph, Success
Danger	Peril, Risk, Hazard	Safety, Security
Deny	Decline, Disown, Refute	Accept, Admit, Confirm
Dirty	Filthy, Impure, Stained	Clean, Pure, Spotless
Dismal	Cheerless, Gloomy, Sad	Cheerful, Jolly, Happy
Divine	Heavenly, Holy, Godly	Earthly, Unholy, Satanic
Dull	Dim, Dry, Stupid, Sad	Bright, Clever, Jovial
Easy	Simple, Convenient	Difficult, Tough
Effective	Active, Operative, Powerful	Ineffective, Futile, Incapable
Efficient	Competent, Capable	Deficient, Incapable, Incompetent
Eminent	Famous, Distinguished	Anonymous, Obscure
Encourage	Inspire, Persuade, Urge	Discourage, Depress, Dissuade
Enemy	Foe, Opponent, Antagonist	Friend, Fellow, Associate
Entire	Full, Complete, Whole	Incomplete, Partial
Essential	Basic, Necessary, Primary	Subsidiary, Secondary, Auxiliary
Eternal	Everlasting, Infinite, Permanent	Transitory, Perishable, Temporary
Evil	Ball, Corrupt, Wicked	Good, Honest, Noble, Moral
Fail	Drop, Miss, Disappoint	Pass, Gain, Attain, Succeed
Fair	Bright, Clean, Proper	Brunette, Dirty, Unfair
False	Deceptive, Illusory, Incorrect	Correct, Real, True
Familiar	Intimate, Accustomed	Strange, Unaccustomed
Famous	Distinguished, Eminent	Notorious, Anonymous
Fast	Agile, Quick, Rapid, Speedy	Dull, Inert, Slack, Slow
Fat	Fleshy, Bulky, Plump	Slim, Thin, Slender, Lean
Fierce	Savage, Wild, Ferocious	Docile, Gentle, Tame
Forward	Progressive, Advanced	Backward, Primitive
Friend	Associate, Ally, Companion	Foe, Enemy, Rival
Happy	Blessed, Cheerful	Unhappy, Sad, Disappointed
Hard	Stiff, Rigid, Solid, Tough	Soft, Flexible, Easy, Mellow
Harmful	Injurious, Mischievous	Harmless, Innocent
Hasten	Accelerate, Quicken	Delay, Retard, hinder
Hate	Abhor, Despise, Dislike	Adore, Love, Like
Heavy	Massive, Weighty	Light, Petty, Trivial

High	Great, Lofty, Tall	Low, Small, Short
Holy	Divine, Godly, Pious	Unholy, Impious, Satanic
Honest	Truthful, Sincere, Frank	Deceitful, Dishonest, Tricky
Honour	Repute, Glory, Respect	Disrespect, Despise, Infamy
Hope	Desire, Expectation, Trust	Despair, Disappointment
Ignorance	Stupidity, Folly, Absurdity	Wisdom, Knowledge, Cleverness
Illegal	Unlawful, Illicit, Forbidden	Legal, Lawful, Legitimate
Important	Significant, Prominent	Unimportant, Trivial
Inactive	Idle, Indolent, Lazy	Alert, Active, Industrious
Incomplete	Unfinished, Deficient Imperfect	Complete, Finished, Perfect
Indolent	Idle, Lethargic, Lazy	Active, Dynamic, Energetic
Inferior	Lower, Subordinate, Mean	Superior, Higher, Greater
Injustice	Unfairness, partiality	Justice, Fairness, Impartiality
Innocent	Honest, Simple, Artless	Cunning, Clever, Tricky
Insult	Abuse, Humiliate, Offend	Admire, Honour, Respect
Intelligent	Bright, Brilliant, Clever	Dull, Foolish, Unintelligent
Irregular	Uneven, Eccentric, Inconsistent	Regular, Systematic, Orderly
Jealousy	Envy, Suspicion, Doubt	Love, Appreciation, Belief
Joy	Ecstasy, Exultation, Pleasure	Grief, Pain, Sorrow, Anguish
Judicious	Wise, Prudent, Careful	Indiscreet, Irrational, Careless
Justice	Fairness, Honesty	Unfairness, Dishonesty
Keen	Eager, Acute, Enthusiastic	Dull, Flat, Lazy
Keep	Detain, Hold, Retain	Abandon, Reject, Discard
Kill	Slay, Murder, Assassinate	Protect, Produce, Restore
Kind	Compassionate, Generous	Cruel, Unkind, Inhuman
Knowledge	Wisdom, Learning	Ignorance, Illiteracy
Known	Familiar, Famous, Popular	Unknown, Ordinary, Simple
Lament	Weep, Mourn, Grieve	Rejoice, Enjoy, Celebrate
Last	Final, Ultimate, Extreme	First, Initial, Foremost
Lavish	Prodigal, Profuse, Excessive	Miserly, Economical, Thrifty
Lawful	Legal, Legitimate, Regular	Unlawful, Illegal, Irregular
Lazy	Idle, Indolent, Slack	Active, Alter, Agile, Quick
Long	Lengthy, Prolonged, Extended	Short, Brief, Compact
Lose	Waste, Fail, Miss	Gain, Obtain, Recover
Loss	Damage, Detriment	Profit, Gain, Earning
Loud	Noisy, Resonant, Blaring	Low, Quiet, Soft
Love	Affection, Fondness, Liking	Hatred, Dislike, Antipathy
Lovely	Attractive, Beautiful, Cute	Ugly, Repulsive, Hideous
Loyal	Faithful, Obedient, Sincere	Disloyal, Treacherous, Unfaithful
Lucky	Fortunate, Successful	Unlucky, Unfortunate
Luxury	Abundance, Plentiful	Scarcity, Paucity, Wanting
Mad	Crazy, Insane, Lunatic	Wise, Sane, Rational
Malice	Hatred, Enmity, Resentment	Love, Passion, Sympathy
Master	Ruler, Chief, Leader	Servant, Slave, Subordinate
Mature	Ripe, Seasoned, Perfect	Raw, Unripe, Immature

Meagre	Inadequate, Scarce, Deficient	Abounding, Ample, Plentiful
Melancholy	Cheerless, Gloomy, Dismal	Cheerful, Gay, Joyful
Merciful	Kind, Gentle, Tender	Cruel, Harsh, Tyrannical
Miserable	Unhappy, Dismal, Poor	Happy, Cheerful, Prosperous
Modern	New, Recent, Progressive	Old, Ancient, Primitive
Natural	Normal, Usual, Regular	Abnormal, Unnatural, Superficial
Necessary	Binding, Essential, Compulsory	Optional, Voluntary, Unnecessary
Neglect	Carelessness, Failure, Omission	Attention, Vigilance, Carefulness
Nice	Fine, Lovely, Pleasing	Repulsive, Ugly, Deformed
Noble	Dignified, Honourable	Ignoble, Mean, Uncultured
Notorious	Scandalous, Ignoble	Famous, Reputable, Virtuous
Numerous	Many, Various, Manifold	Few, Scarce, Scanty
Obnoxious	Odious, Offensive, Perverse	Pleasant, Enjoyable, Nice
Obscene	Dirty, Lewd, Filthy	Pure, Virtuous, Clean
Obstinate	Stubborn, Adamant, Resolute	Amenable, Yielding, Accommodative
Occasional	Sporadic, Casual, Incidental	Frequent, Regular, Repeated
Odd	Abnormal, Anomalous, Singular	Normal, Customary, Even
Offend	Hurt, Displease, Provoke	Defend, Please, Flatter
Often	Usually, Repeatedly, Generally	Rarely, Hardly, Seldom
Old	Aged, Primitive, Elderly	Young, Modern, Immature
Ominous	Inauspicious, Menacing	Auspicious, Propitious
Omit	Delete Drop, Eliminate	Include, Insert, Enroll
Optimistic	Hopeful, Promising, Inspiring	Hopeless, Disheartening, Pessimistic
Ordinary	Normal, Regular, Usual	Abnormal, Exceptional, Unusual
Outstanding	Eminent, Important, Excellent	Unimportant, Ordinary, Normal
Pain	Agony, Anguish, Pang	Joy, Pleasure, Happiness
Passionate	Excited, Hasty, Impulsive	Dispassionate, Cool, Balanced
Pathetic	Pitiable, Sad, Touching	Amusing, Comic, Farcical
Peace	Tranquility, Amity, Silence	War, Turmoil, Upheaval
Pious	Religious, Devout, Godly	Impious, Irreligious, Sinful
Pleasant	Attractive, Nice, Amiable	Repulsive, Unpleasant, Harsh
Poison	Venom, Toxin, Virus	Nectar, Honey, Antidote, Sweet
Polite	Courteous, Mannerly, Cordial	Impolite, Offensive, Rude
Positive	Affirmative, Categorical, Sure	Negative, Vague, Dubious
Powerful	Energetic, Strong, Mighty	Powerless, Weak, Dull, Frail
Precious	Valuable, Expensive, Esteemed	Worthless, Cheap, Contemptible
Profit	Advantage, Privilege, Gain	Loss, Damage, Deprivation
Prohibit	Debar, Forbid, Stop, Hinder	Permit, Authorise, Sanction
Protect	Defend, Guard, Shield	Expose, Disclose, Reveal
Quantity	Bulk, Measure, Amount	Quality, Deficit, Shortage
Quarrel	Dispute, Fight, Feud	Peace, Harmony, Amity
Question	Inquiry, Interrogation, Query	Answer, Reply, Solution
Questionable	Doubtful, Debatable, Uncertain	Assured, Certain, Unquestionable
Quick	Alert, Swift, Fast, Rapid	Slow, Slack, Lethargic, Sluggish

Word	Synonyms	Antonyms
Rapid	Fast, Quick, Speedy, Swift	Slow, Sluggish, Indolent
Rare	Unique, Sparse, Exclusive	Common, Usual, Ordinary
Rational	Judicious, Sensible, Normal	Irrational, Erratic, Insane
Reasonable	Fair, judicious, Sensible	Illogical, Absurd, Ridiculous
Religious	Holy, Devout, Pious, Saintly	Unholy, Sinful, Impious, Irreligious
Remember	Recollect, Recall, Retain	Forget, Ignore, Overlook
Respectable	Honourable, Reputable, Worthy	Contemptible, Disreputable, Unworthy
Rich	Wealthy, Abundant, Profuse	Poor, Scarce, Inadequate
Right	Correct, Exact, Proper, True	Wrong, False, Unfair, Bad
Rough	Hard, Coarse, Crude	Soft, Smooth, Decent
Rude	Impolite, Rough, Mannerless	Courteous, Refined, Mannerly
Rural	Rustic, Agrarian, Pastoral	Urban, Civic, Metropolitan
Ruthless	Merciless, Cruel, Pitiless	Kind, Compassionate, Pitiful
Sacred	Holy, Religious, Divine, Pious	Blasphemous, Unholy, Sinful
Safe	Protected, Secure	Exposed, Unsafe
Scarcity	Dearth, Paucity, Shortage	Prosperity, Abundance, Enormity
Selfish	Greedy, Mean	Generous, Liberal, Charitable
Sensitive	Tender, Delicate, Susceptible	Insensitive, Blunt, Robust
Shallow	Superficial, Simple, Trivial	Deep, Profound, Wise
Sharp	Edged, Alert, Keen	Blunt, Dull, Flat
Strict	Austere, Firm, Stern	Mild, Flexible, Lenient
Superior	Excellent, Higher, Finer	Inferior, Unimportant, Mean
Tactful	Diplomatic, Skilful, Prudent	Simple, Negligent, Stupid
Temporary	Interim, Ephemeral	Permanent, Imperishable
Tender	Delicate, Soft, Fragile	Hard, Firm, Rigid
Tentative	Provisional, Experimental	Final, Definite, Firm
Thick	Dense, Solid, Condence	Thin, Slim, Slender
Thin	Delicate, Slim, Slender	Thick, Solid, Dense
Tough	Sturdy, Tenacious, Firm	Flexible, Soft, Tender, Weak
Transparent	Clear, Distinct, Glassy	Opaque, Hazy, Unclear
Treacherous	Disloyal, Deceitful, Unfaithful	Loyal, Faithful, Devoted
Ugly	Repulsive, Hideous, Vicious	Pretty, Beautiful, Charming
Unanimity	Agreement, Harmony, Unison	Conflict, Discord, Differences
Uncertain	Indefinite, Doubtful, Dubious	Definite, Certain, Sure
Uneasy	Disturbed, Restless, Worried	Steady, Quiet, Restful
Unfair	Unjust, Wrongful, Biased	Just, Fair, Impartial
Unfaithful	Disloyal, Deceitful, Treacherous	Faithful, Reliable, Loyal
Unfit	Improper, Incompetent, Unsuitable	Fit, Capable, Competent, Suitable
Unfortunate	Unlucky, Ruined, Unfavoured	Blessed, Fortunate, Lucky
Unholy	Impious, Sinful, Irreligious	Holy, Pious, Religious
Unique	Uncommon, Matchless, Rare	Normal, Ordinary, Common
Unity	Oneness, Harmony, Union	Diversity, Difference, Variance
Vacant	Empty, Unfilled, Void	Occupied, Packed, Inhabited
Valiant	Brave, Gallant, Heroic	Timid, Cowardly, Frightened
Vanity	Pride, Arrogance, Egotism	Modesty, Humbleness, Humility
Vast	Huge, Enormous, Boundless	Small, Limited, Scanty
Vertical	Erect, Perpendicular, Upright	Horizontal, Flat, Prone

Vice	Sin, Immorality, Corruption	Virtue, Morality, Honesty
Victory	Conquest, Success, Triumph	Defeat, Failure, Downfall
Vigilant	Alert, Cautious, Watchful	Careless, Negligent, Slack
Violent	Fierce, Furious, Passionate	Unexcited, Quite, Cool
Virtue	Morality, Purity, Holiness	Vice, Wickedness, Sin
War	Battle, Combat, Conflict	Peace, Harmony, Amity
Warmth	Cordiality, Earnestness	Apathy, Indifference, Coolness
Wary	Alert, Cautions, Prudent	Careless, Negligent, Rash
Watchful	Attentive, Alert, Careful	Careless, Negligent, Uncautious
Weak	Feeble, Fragile, Poor	Strong, Sturdy, Powerful
Whole	Complete, Entire, Total	Part, Fractional, Incomplete
Wicked	Evil, Sinister, Villainous	Noble, Virtuous, Chaste
Wisdom	Intelligence, Cleverness	Absurdity, Stupidity, Folly
Yearn	Crave, Desire, Languish	Contented, Satisfied, Cheerful
Young	Youthful, Fresh, New	Old, Mature, Ripe
Zeal	Enthusiasm, Passion, Fervour	Apathy, Reluctance, Idleness
Zenith	Climax, Apex, Crown, Culmination	base, Bottom, Foundation, Minimum

(C) HOMOPHONES/HOMONYMS

In the English language, there are a number of words which are spelt and pronounced alike but carry different meanings. Such words are called as *homonyms*. Similarly, there are words pronounced like another words but with different meaning or spelling. Those words are called as *homophones*. These peculiarities of English words having great resemblance in spelling or pronunciation but differences in their meaning, create confusion in the minds of students lead to committing mistakes in their usages. Some of the words belonging to these categories are given below indicating their different meanings and using them in appropriate sentences bringing home those different shades of meanings.

(i) **Words with the same spelling and pronunciation but carrying different meaning. (Homonyms).**

   1. **Act**
   (a) Decree on law made by a legislative body.

   Ex. Government have passed an act to prohibit the sale of drugs.
   (b) Any of the main divisions of a play or an opera.

   Ex. The tempo of the play culminates in the III Act.

   2. **Agent**
   (a) Person who acts for or manages the affairs of other people in business, politics etc.

   Ex. He works as an insurance agent.
   (b) Force or Substance that produces an effect or change.

   Ex. Yeast is used these days in many recipies as the raising agent.

3. Bar

(a) A long shaped piece of hard stiff material.

Ex. A large number of metal bars were piled up in the godown.

(b) A counter at which drinks are served.

Ex. They were found in the bar in intoxicated condition

4. Bat

(a) Small mouse-like animal that flies at night and feeds on fruits and insects.

Ex. He found bats flying in the house that remained closed for years.

(b) A wooden implement of a specified size and shape and with a handle used for hitting the ball in the games like cricket, baseball and table tennis.

Ex. He used a little heavier bat during the tournament.

5. Beat

(a) Stroke or regular sequence of strokes.

Ex. We heard the drum beat early in the morning.

(b) Emphasis repeated regularly marking rhythm in music or poetry.

Ex. That song has a very good and melodious beat.

6. Capital

(a) Wealth or property that may be used for producing more wealth.

Ex. He started his business with a meagre amount of capital.

(b) Town or City that is the centre of the Government of a country.

Ex. Mumbai is the capital of the Maharashtra State.

7. Card

(a) Thick stiff paper on thin pasteboard.

Ex. Identity card is compulsory for all college students.

(b) Wire brush or toothed instrument for cleaning or combing wool.

Ex. He used his card very carefully and efficiently in the whole department.

8. Date

(a) Specific numbered day of the month or specific year usually given to show when something happened or is to happen.

Ex. The date of the conference collided with the date of his birth.

(b) Brown sweet edible fruit of a palm-tree common in N. Africa and S. W. Asia.

Ex. Date is a useful fruit for health.

9. Drill

(a) Training in military exercises.

Ex. N. C. C. cadets have three hours of drill everyday.

(b) Machine for making furrows, sowing seeds in them and covering the seeds by soil.

Ex. He purchased a new drill for his farming.

10. Even

(a) Level, smooth, flat.

Ex. Highways should be perfectly even for smooth traffic.

(b) Divisible by two with no remainder.

Ex. The teacher taught today the sum of even numbers.

11. Fan

(a) Device with rotating blades operated mechanically to create a current of cool air.

Ex. We installed a new ceiling fan in our drawing room.

(b) Enthusiastic admirer or supporter.

Ex. A lot of people are cinema fans.

12. Fence

(a) Structure of wire etc. put round a field, or a garden to make a boundary or keep animals from straying.

Ex. It was necessary to put a fence round our garden.

(b) Person who knowingly buys and resells stolen goods.

Ex. He was arrested on the charge of being a fence.

13. Gin

(a) Trap or share for catching animals.

Ex. The hunter was aware where the gin was placed.

(b) Colourless alcoholic drink often used in cocktails.

Ex. He preferred gin with water as a tonic.

14. Gloss

(a) Brightness or shine on a smooth surface.

Ex. Heavy polishing was necessary to bring gloss on the old wooden furniture.

(b) Explanatory comment; explanation, interpretation.

Ex. The election candidate put a different gloss on the issue of corruption in the political field.

15. Hip

(a) Part on either side of the body below the waist where the bone of a person's leg is jointed to the trunk.

Ex. He was standing there with his hands on his hips.

(b) Berry-like fruit of the wild rose.

Ex. Did you see hips on the fence of our garden ?

16. Jade

(a) Hard, usually green, stone from which ornaments are carved.

Ex. He owned a beautiful jade vessel.

(b) Tired or worn-out horse.

Ex. He came across a jade on his way back home.

17. Labour

(a) Physical or mental work.

Ex. Workers naturally expect appropriate money wage for their labour.

(b) Workers as a group or class.

Ex. The factory engaged a large number of unskilled labour last month.

18. Law

(a) Rule established by authority.

Ex. The old law has already become obsolete.

(b) Subject of study.

Ex. He took up law after his first graduation in commerce.

19. Magazine

(a) Paper-covered periodical, usually weekly or monthly.

Ex. He purchased a magazine for reading during his long journey by train.

(b) Store of arms, ammunition, explosives etc.

Ex. He enquired about the prices of guns in the magazine on his way home.

20. Margin
(a) Blank space round the written or printed matter.
    Ex. The teacher directed to leave a sufficient margin before beginning to write the paper.
(b) Difference between cost-price and selling-price.
    Ex. He discarded the deal because there was hardly any profit margin.

21. Mess
(a) Dirty or untidy state.
    Ex. Spectators made mess in the auditorium on a lot off failure of the electricity.
(b) Building in which people take meals together.
    Ex. It is the best mess in our locality.

22. Nature
(a) The whole universe and every creation, not a man-made, thing.
    Ex. The wonders of nature are beyond human imagination.
(b) Typical qualities and characteristics of a person or animal.
    Ex. Corruption is not in his nature.

23. Negative
(a) Word or statement that expresses or means denial or refusal.
    Ex. The words like *no, not, neither* are negatives.
(b) Developed photographic film from which photoprints are taken.
    Ex. All the negatives of our tour to Simla are spoiled by the photographer.

24. Order
(a) Way in which people or things are arranged in relation to one another.
    Ex. The names were arranged in the alphabetical order.
(b) Command or instructions given by somebody in authority.
    Ex. The cadets obeyed the orders given by the commander.

25. Part
(a) Distinct portion of a human or animal body or of a plant.
    Ex. Which part of your body is hurt ?
(b) Role played by an actor in a play, film etc.
    Ex. His rolicking part in the film was appreciated by the public.

26. Play
(a) Activity done for amusement esp. by children.
    Ex. The park was full of the happy sounds of children at play.
(b) Drama
    Ex. The new play staged yesterday for the first time was quite excellent.

27. Pole
(a) Either of the two ends of a magnet or the terminal points of an electric battery; either of the two points at the exact-top or bottom of the earth.
    Ex. The arctic is the region around the North Pole.
(b) Long thin rounded piece of wood or metal used esp. as a support for something or for pushing a boat etc. along.
    Ex. He is expert at climbing the telegraph pole.

28. Raw
(a) Uncooked
    Ex. Eat vegetables like carrot, radish etc. raw for good health.

(b) Inexperienced, unskilled.

    Ex. Raw hands would not help to improve the conditions of the company.

29. Right

(a) Best in view of the circumstances, most suitable.

    Ex. We are on the right track of action.

(b) Contrasted with left.

    Ex. She was standing by my right-hand side.

30. Saw

(a) A long cutting tool with sharp toothed edge.

    Ex. He installed a new saw in his factory.

(b) Saying, Proverb.

    Ex. The old saw 'A stitch in time saves nine'.

31. Seal

(a) Animal with flippers that lives near and in the sea and eats fish.

    Ex. A number of seals were seen on the beach.

(b) Soft material such as wax etc. stamped with a design and fixed to a document to show that it is genuine and to prevent it to be opened by a wrong person.

    Ex. The envelop was affixed with the seal of the king.

32. Second

(a) Person or thing that comes next after the first.

    Ex. He stood second in the elocution competition.

(b) 60th part of a minute of time.

    Ex. The state of the game was changed within five seconds.

33. Share

(a) Part or portion of a larger amount divided among many people or to which many people contribute.

    Ex. Everyone present will get a fair share of the food.

(b) Any of the equal parts into which the capital of a business company is divided.

    Ex. He held 500 shares of that co-operative credit bank.

34. Sight

(a) Ability to see, vision.

    Ex. He lost his sight in the recent accident.

(b) Range within which somebody can see or something can be seen.

    Ex. In few seconds, his car was out of our sight.

35. Slip

(a) Act of Slipping; false step.

    Ex. There is many a slip between the cup and the lip.

(b) A thing or small piece of paper.

    Ex. I noted down the phone number on a slip of paper.

36. Spirit

(a) Soul thought of as separate from the body; soul without a body; ghost.

    Ex. People say that the house on the other side of the road is haunted by an evil spirit.

(b) Strong, distilled alcoholic drink.

    Ex. My friend is habituated to drink spirits like rum, whisky, brandy etc.

37. Tank

(a) A large container generally for liquid or gas.

Ex. His business is to give tanks on hire for transport of milk.

(b) Armoured fighting vehicle with guns which moves on caterpillar tracks.

Ex. A large number of tanks of the enemy were destroyed in the battle.

38. View

(a) What can be seen from a particular place; fine natural scenery.

Ex. You will get a better view of the valley when looking from a railway window.

(b) Personal opinion or attitude.

Ex. His views about today's political situation are very bitter.

39. Ward

(a) Separated part or room for a particular group of patients in a dispensary or hospital.

Ex. You can meet him in the surgical ward.

(b) Division of a city that elects and is represented by a councillor in the local government.

Ex. The councillor from our ward is chosen as a mayor of the city.

40. Wave

(a) Move regularly and loosely to and fro or up and down.

Ex. Our tricolour was waving proudly in the air.

(b) Move one's hand to and fro or up and down in order to attract attention, or to make a signal or give a greeting.

Ex. All the family members waved their hands as the train moved from the station.

Paronyms :

Paronyms are words of the same grammatical class (nouns, adjectives, etc.) that have the same root as another or a word containing the same root as another. A word which is borrowed or descended from the same origin.

For example,

–   wise and wisdom.

The word wisdom has the same root as wise. Hence, a word from the same root or having the same sound as that in another can be taken as Paronymouns.

Example :

social	–	society
ornate	–	ornateness
orgon	–	organic
beauty	–	beautiful
Book	–	Bookish
style	–	stylish

(ii) **Words with same pronunciation but different-spelling or meaning are called as *homophones*. Some of the examples of such words are given below.**

1. **Advice - Advise.**

(a) **Advice -** Opinion given about what to do, or how to behave.

Ex. His advice to his friend proved to be beneficial.

(b) **Advise -** Recommend, give advice to somebody.

Ex. The doctor advised him to take complete rest.

2. **Aid-Aide.**

(a) **Aid -** Help, food, money etc. sent to a country to help it.

Ex. How much foreign aid did we receive during the recent floods ?

(b) Aide - Assistant.

Ex. He is one of the chief aides to the President.

3. Beat - Beet

(a) Beat - Hit something repeatedly especially with a stick

Ex. Somebody was beating at the door.

(b) Beet - Type of plant with a fleshy root which is used as a vegetable or for making sugar.

Ex. Beet is a useful vegetable for health reasons.

4. Canvas - Canvass.

(a) Canvas - Strong coarse cloth used for making tents etc. and by artists for painting on.

Ex. He purchased a canvas for his painting work.

(b) Canvass - Go around an area for political support.

Ex. He participated in the canvassing campaign.

5. Cell - Sell

(a) Cell - A very small room; microscopic unit of living matter.

Ex. The prisoners were kept in a single cell in the basement.

(b) Sell - To give goods etc. to somebody who pays money.

Ex. He goes to market early in the morning to sell books.

6. Dear - Deer

(a) Dear - Greatly valued, loved; expensive.

Ex. His daughter was very dear to him.

(b) Deer - Any of the several types of graceful, quick-running animal.

Ex. There were many beautiful deers in the zoo.

7. Die - Dye

(a) Die - Stop living, come to the end of one's life.

Ex. His relative died of a heart attack.

(b) Dye - Substance used for colouring.

Ex. He constantly uses a hair dye to look young.

8. Desert - Dessert

(a) Desert - Barren land with very little water and vegetation, often covered by sand.

Ex. The Sahara Desert is the largest one in the world.

(b) Dessert - Any sweet dish eaten at the end of a meal (e.g. ice-cream, fruit-salad etc.).

Ex. We shall now move on to dessert, if the meals are over.

9. Fair - Fare

(a) Fair - Large scale exhibition of commercial and industrial goods.

Ex. A trade fair was recently held in the capital.

(b) Fare - Money charged for a journey by bus, taxi, ship etc.

Ex. The bus fare has exorbitantly increased.

10. Gait - Gate

(a) Gait - Manner of walking or running.

Ex. He was walking with an unsteady gait.

(b) Gate - Means of entrance or exit; movable barrier which closes on opening in a wall, fence etc.

Ex. The garden gate is completely broken and needs repairs.

11. **Hair - Hare**

(a) **Hair -** Mass of the fine thread-like strands that grow from the skin of people and animals, esp. on the human head.

    Ex. She has beautiful long, black hair which attracts people.

(b) **Hare -** Fast running rabbit-like but larger mammal with long ears and divided upper lips, that lives in fields.

    Ex. Hares are very rarely seen even in the forests.

12. **Hall - Haul**

(a) **Hall -** Building or large room for meetings, meals etc.

    Ex. The meeting was convened in the town hall.

(b) **Haul -** Pull or drag something with effort.

    Ex. Sailors hauled the boat up the beach.

13. **Idle - Idol**

(a) **Idle -** Doing or having no work; not employed.

    Ex. Idle people cause nuisance to the whole society.

(b) **Idol -** Person or thing that is greatly loved or admired.

    Ex. Actors or actresses are always the matinee idols for people.

14. **Knight - Night**

(a) **Knight -** Man to whom the sovereign has given a rank of honour having the title of 'Sir'. Ex. Rabindranath Tagore had refused the knightship honoured to him by the British sovereign.

(b) **Night -** Time of darkness between sunset and sunrise.

    Ex. People generally take rest during the night.

15. **Main-Mane**

(a) **Main -** Most important, chief; principal.

    Ex. She was the main person of attraction in the party.

(b) **Mane -** Long hair on the neck of a horse, lion etc.

    Ex. Lions look graceful because of the mane they possess.

16. **Manner-Manor**

(a) **Manner -** Way in which a thing is done or happens.

    Ex. The manner in which he behaved was not graceful.

(b) **Manor -** Unit of land under the feudal system, part of which was used by the lord and the rest being farmed by his tenants.

    Ex. Tenants in the manor were not always treated well by their lords.

17. **Not - Nought**

(a) **Not -** Used with auxiliary verbs and modals to form the negative.

    Ex. He did not see me during the whole of last month.

(b) **Nought -** The figure zero (0); nothing.

    Ex. Put three noughts after the figure 5.

18. **One - Won**

(a) **One -** One less than two, single.

    Ex. He was the only one present in the hall.

(b) **Won -** Past tense and past participle form of the verb win.

    Ex. India won the Hero trophy match by a narrow margin.

19. Peer - Pier

(a) Peer - Person who is equal to another in rank, status or merit.

Ex. He likes to mix up only with his peers.

(b) Pier - Structure built out into the sea, lake etc. so that boats can stop and take on or put down goods or passengers.

Ex. He was waiting at the pier for a boat to arrive.

20. Quote - Cote

(a) Quote - Repeat in speech or writing words previously said or written by another person.

Ex. He always quotes lines from Sanskrit verses.

(b) Cote - Shed, shelter or enclosure for domestic animals or birds.

Ex. All the chicken were enclosed in a cote during the night.

21. Right - Rite

(a) Right - Morally good conduct, action; required by law or duty.

Ex. His reaction on the issue was quite right.

(b) Rite - Religious or some other solemn ceremony.

Ex. He took leave for performing certain religious rites at home.

22. Sew - Sow

(a) Sew - Make stitches in cloth etc. with a needle and thread.

Ex. She is sewing her gown since morning.

(b) Sow - Put or scatter seeds in or on the ground.

Ex. The farmer sowed cabbage seeds in the farm.

23. Sight - Site

(a) Sight - Ability to see, vision.

Ex. Due to old age his sight has become a little weak.

(b) Site - Place where a building, town etc. was, is or will be situated;

24. Tail - Tale

(a) Tail - Movable part at the end of the body of a bird, an animal, a fish or reptile.

Ex. Dogs wag their tails when they are in a happy mood.

(b) Tale - Narrative or story.

Ex. Children really like fairy tales.

25. Urn - Earns

(a) Urn - A tall vase usually with a stem and a base; Large metal container with a tap for making or serving tea, coffee etc. in hotels.

Ex. The urn contained a large quantity of coffee.

(b) Earn - Get money etc. by working.

Ex. He earned his wages once in a fortnight.

(D) SAME WORD USED AS DIFFERENT PARTS OF SPEECH

Sometimes we get derivatives from a root word or headword without adding a prefix or suffix. However, such derivatives carry different meaning and belong to the part of speech other than the one to which the original word belongs. Examples of some such zero derivatives are given below.

1. **Advance (n) :** Forward movement.
   Ex. His advance in life was obstructed because of his financial position.
   Advance (adj) : Going before others, done or provided in advance.
   Ex. He was a given an advance notice about his dismissal.

2. **Advocate (v) :** Speak publicly in favour of something or recommend.
   Ex. He always advocated the Presidential type of government.
   Advocate (n) : Person who supports or speaks in favour of a cause or policy.
   Ex. Nehru was lifelong advocate of the 'Panchsheel'.

3. **Bale (n) :** Large bundle of paper, straw, goods etc. pressed together and tied with rope or wire for transport etc.
   Ex. The bales of hay in the farm caught fire.
   Bale (v) : Make something into or pack in bales.
   Ex. The attendant was busy in baling cut-pieces of cloth.

4. **Balloon (n) :** Brightly coloured rubber bag that is filled with air used as a child's toy or a decoration.
   Ex. A number of different coloured balloons were brought home for celebrating child's birthday.
   Balloon (v) : Swell out like a balloon.
   Ex. His loose attire ballooned in the heavy wind.

5. **Cable (n) :** Set of insulated wires for carrying messages by telegraph, or electricity overhead, message sent abroad.
   Ex. Company used copper cables on the new lines.
   Cable (v) : Send a cable to somebody abroad.
   Ex. His success was cabled to him without losing any time.

6. **Camp (n) :** Place where people live temporarily in tents or huts.
   Ex. The party pitched its camp by the river bank.
   Camp (v) : Put up a tent or tents.
   Ex. Our plan was to camp in the village for a night.

7. **Ditch (n) :** Narrow channel dug at the edge of a field road etc. esp. to hold or carry water.
   Ex. That ditch proved to be dangerous for the pedestrians.
   Ditch (v) : Land an aircraft in the sea in an emergency.
   Ex. The pilot was forced to ditch in the Arabian sea due to defect in the engine.

8. **Documentary (adj) :** Consisting of documents (documentary evidence, proof, sources).
   Ex. The file had a documentary, important to the court.
   Documentary (n) : A documentary film, or radio or T. V. programme.
   Ex. His documentary received critics' award.

9. **Drizzle (v) :** To rain in many fine drops.
   Ex. It has been drizzling all day.

Drizzle (n) : Fine misty rain.

Ex. There was a cold drizzle since morning.

10. **Earth (n) :** This planet, the planet on which we live; soil.

Ex. The earth rotates round the sun.

Earth (v) : Connect an electrical appliance etc. with the ground.

Ex. The connection of the electric iron was earthed.

11. **Eccentric (adj) :** Unusual, peculiar, not normal.

Ex. An eccentric person next door is a nuisance to all.

Eccentric (n) : Abnormal person.

Ex. That group is overflowing with eccentrics.

12. **Flight (n) :** Process or action of flying in the air.

Ex. The aeroplane was shot down in flight.

Flight (v) : Give a certain patch to a ball through the air while bowling so as to deceive the batsman.

Ex. He made a well flighted delivery of his last ball of the over.

13. **Flirt (v) :** To behave in a romantic way but without serious intentions.

Ex. He always tries to flirt with girls in his office.

Flirt (n) : A person who flirts with many people.

Ex. He is known as a terrible flirt in the whole office.

14. **General (adj) :** Affecting all or most people, places or things.

Ex. The proposal received a general approval of the people.

General (n) : An army officer of a very high rank.

Ex. He retired as a general from the Indian Army.

15. **Gesture (n) :** Expressive movement particularly of a hand or head.

Ex. He tried to communicate his desire through gestures.

Gesture (v) : Make expressive movements; convey something by gestures.

Ex. She gestured her happiness on the proposal of her friend.

16. **Handicap (n) :** A thing that makes progress difficult; physical or mental disability.

Ex. Blindness is beyond doubt a serious handicap.

Handicap (v) : To give or be a disadvantage.

Ex. He was handicapped due to his scanty income.

17. **Handle (n) :** A part of a tool, cup, bucket, drawer etc. by which it can be held, carried or controlled.

Ex. His car cannot start without using a handle.

Handle (v) : Deal with, manage or control (people, a situation, a machine etc.)

Ex. The foreman very well knew how to handle the workers under him.

18. **Independent (adj) :** Not dependent (on other people or things); not controlled (by other people or things).

Ex. India was once a British colony but now it is independent.

Independent (n) : MP, candidate etc. who does not belong to any political party.

Ex. He filed his nomination as an independent to contest election.

19. **Individual (adj) :** Single, separate.

Ex. Each individual is responsible for his own deeds.

Individual (n) : A single human being.

Ex. The rights of an individual are controlled in the dictatorship.

20. **Jumble (v)** : Mix things in a confused way.
    Ex. Everything in the hall was jumbled.
    Jumbled (n) : Untidy group of things; muddle.
    Ex. It was a jumble of books on the table.

21. **Key (n)** : Metal instrument for locking or unlocking.
    Ex. He lost the key of his door-latch.
    Key (adj) : Very important or essential.
    Ex. He is the key person in his company.

22. **Lament (v)** : Feel or express great sorrow or regret.
    Ex. They lamented upon the loss of their popular leader.
    Lament (n) : Strong expression of grief.
    Ex. People were greatly moved by her laments on the loss of her only child.

23. **Large (adj)** : Of considerable size, extent or capacity.
    Ex. A large country needs large sources of production.
    Large (n) : At full length, thoroughly and in great details.
    Ex. The matter is discussed at large in the report.

24. **Minor (adj)** : Smaller, less serious, less important.
    Ex. The house requires, minor repairs.
    Minor (n) : A person under the age of full legal responsibility.
    Ex. He is not eligible to vote as he is a minor.

25. **Misuse (v)** : Not used properly.
    Ex. Use in the wrong way or for the wrong purpose.
    Ex. Nobody should misuse the public funds.
    Misuse (n) : You should always avoid misuse of power in public life.

26. **Nail (n)** : Layer of thorny substance over the outer tip of a finger or toe; a small thin piece of metal.
    Ex. She defended herself with her strong pointed nails when the stranger tried to molest her.
    Nail (v) : Catch or arrest.
    Ex. The police nailed the culprit within 24 hours of the offence.

27. **National (adj)** : Of a nation; characteristic of a whole nation.
    Ex. There is always difference in the local and national newspapers.
    National (n) : A citizen of a particular nation.
    Ex. Only Indian nationals are allowed to contest election.

28. **Objective (adj)** : Unbiased; fair; not influenced by personal feelings or opinions.
    Ex. The magistrate's decision in the case was perfectly objective.
    Objective (n) : Thing aimed at or wished for; purpose; aim.
    Ex. To become a lawyer was his sole objective.

29. **Patent (n)** : Official document giving the holder the sole right to make, use or sell an invention and preventing others from imitating it.
    Ex. He has already applied to the government for a patent of the gadget he invented very recently.
    Patent (v) : Obtain a patent for an invention or process.
    Ex. He was patented for his invention.

30. **Poison (n)** : Substance causing death or harm if absorbed by living thing (animal or plant).
    Ex. The actor committed suicide by taking poison.

Poison (v) : To give poison to living thing; kill or harm something/somebody with poison.

Ex. She poisoned her husband for getting his wealth.

31. **Quarter (n) :** Each of four equal or corresponding parts of something.

    Ex. Only one quarter of the theatre was empty.

    Quarter (v) : Provide somebody with lodging.

    Ex. We were quartered in the government rest houses on payment.

32. **Refill (v) :** To fill again.

    Ex. He had to refill the petrol tank before starting the journey.

    Refill (n) : New material used to refill a container.

    Ex. Bring two refills for my ballpen while coming home.

33. **Reform (v) :** Become or make better by removing faults, errors etc.

    Ex. Many social organizations desire to reform the structure of our constitution.

    Reform (n) : Reforming or being reformed.

    Ex. They plan to bring about social reforms quickly.

34. **Screen (n) :** Blank surface on which pictures or films are projected.

    Ex. He writes for both the big and small screen.

    Screen (v) : show a film, scene on a screen.

    Ex. The film has been screened in the cinema and on T.V.

35. **Sheer (adj) :** Complete, thorough, utter.

    Ex. His efforts were a sheer waste of time.

    Sheer (adv) : Straight up or down.

    Ex. The ground dropped, and it was her sheer luck that saved her.

36. **Tape (n) :** Magnetic tape on which recording is made.

    Ex. The police could not finish the tape they needed.

    Tape (v) : Record on magnetic tape.

    Ex. The concert was taped from the T. V.

37. **Thrust (v) :** Push violently or suddenly.

    Ex. He thrust the dagger through the body of the thief.

    Thrust (n) : Act or movement of thrusting.

    Ex. The soldier was killed by a bayonet thrust.

38. **Unlike (adj) :** Different, dissimilar.

    Ex. They are quite unlike each other.

    Unlike (prep) : Not like; different from.

    Ex. His performance was quite unlike the earlier one.

39. **Venture (n) :** Project or undertaking involving a risk of failure.

    Ex. His friend embarked on a doubtful venture.

    Venture (v) : Dare to go to do something dangerous.

    Ex. The mouse never ventured far from its hole.

40. **Yelp (n) :** A short, sharp cry of pain, anger, excitement etc.

    Ex. The dog gave a yelp when the boy trodded on its paw.

    Yelp (v) : To give a short, sharp cry of pain, anger etc.

    Ex. She yelped with joy when she saw her friend standing in front of her.

(4) USE OF CONTEXTUAL WORDS IN A GIVEN PARAGRAPH

This topic is a double-edged weapon in the hands of the students since it helps both to elaborate ideas, concepts, thoughts for writing as well as to bring precision when required to write or express things in short. Although it is impossible to list out all one-words, efforts are made to compile some important and commonly used groups of words and one-word substitutes for them in this section. Students are advised to study them and try to remember them by using them in practice whenever and wherever possible. This will also be one of the measures for enriching one's vocabulary.

1. People inhabiting a land from a very early period before the colonies were established.
Aboriginals

2. Something that exists in thought or idea but not having a physical or practical existence.
Abstract

3. A person whose profession is to maintain or inspect financial accounts. **Accountant**

4. A word formed from the initial letters of a group of words. **Acronym**

5. Concerned with beauty and the appreciation of beauty. **Aesthetic**

6. Instance of unprovoked attacking or hostility by one country against another. **Aggression**

7. Medical conditions that produce an unfavourable reaction to certain food, pollens, insect bites, etc. **Allergy**

8. A diplomat sent from one country to another either as a permanent representative or on a special mission. **Ambassador**

9. Supply of bullets, bombs, grenades etc. fired from weapons or thrown. **Ammunition**

10. An animal able to live both on land and in water. **Amphibian**

11. Conditions of being lawlessness i.e. the complete absence of government law or control of society. **Anarchy**

12. Scientific study of the structure of animal bodies. **Anatomy**

13. With a name that is not known or not made public. **Anonymous**

14. Study of mankind esp. of its origin, development, customs and beliefs. **Anthropology**

15. A word that is opposite in meaning to another. **Antonym**

16. An artificial pond or glass tank where live fish and other water creatures and plants are kept. **Aquarium**

17. Study of ancient civilizations by scientific analysis of physical remains found in the ground. **Archaeology**

18. Art and science of designing and constructing buildings. **Architecture**

19. Part of a country's military forces that is organized and equipped for fighting on land.
Army

20. Study of the positions of the stars and movements of planets in the belief that they influence human affairs. **Astrology**

21. Person who travels in a space-craft. **Astronaut**

22. Scientific study of the sun, moon, stars, planets etc. **Astronomy**

23. Story of a person's life written by that person himself. **Autobiography**

24. Microscopic organisms existing in air, water, soil, living and dead creatures, plants causing diseases. **Bacteria**

25. Style of dancing used to tell a story in a dramatic performance with music but without speech or singing. **Ballet**

26. A strip of material used for binding round a wound or an injury. **Bandage**

27. An instrument for measuring atmospheric pressure especially used for forecasting weather. **Barometer**

28. A lawyer who has acquired the degree of Bar-at-Law and has right to speak and argue as an advocate in higher law courts. **Barrister**

29. Stretch of sand or pebbles along the edge of the sea or a lake; shore between high and low water mark. **Beach**

30. Person who gives money or other help to a school, hospital, charitable institute etc. **Benefactor**

31. Being or wishing to be kind, friendly and helpful to or towards somebody. **Benevolent**

32. List of books or articles about a particular subject or by a particular author. **Bibliography**

33. System or crime of marrying a person when still legally married to someone else. **Bigamy**

34. Having two sides; affecting or involving two parties, countries etc. **Bilateral**

35. Able to speak two languages equally well. **Bilingual**

36. Scientific study of plants and their structure. **Botany**

37. Structure of wood, iron, concrete etc. providing a way across a river road, railway etc. **Bridge**

38. Meal at which guests serve themselves from a number of dishes. **Buffet**

39. Powerful tractor with a broad steel blade in front used for moving earth or clearing ground. **Bulldozer**

40. System of government through departments managed by state officials not by elected representatives. **Bureaucracy**

41. Piece of furniture with drawers or shelves for storing or displaying things; **Cabinet**

OR

Group of the most important government ministers responsible for government administration and policies.

42. Person whose job is to receive and pay out money in bank, shop, hotel counter etc. **Cashier**

43. Small sealed case containing a reel of film or magnetic tape. **Cassette**

44. Book or a booklet containing a complete list of items usually in a special order and with a description of each item. **Catalogue**

45. Place serving food and drink in a factory, an office, a school, etc. **Canteen**

46. Economic system in which a country's trade and industry are controlled by private owners and not by state for profit. **Capitalism**

47. Branch of medicine concerned with the heart and its diseases. **Cardiology**

48. Branch of mathematics that deals with the problems involving rates of variation. **Calculus**

49. Strong smelling white substance used in medicines and mothballs and making plastics. **Camphor**

50. Official counting of a country's population or of other classes of things for statistical purposes. **Census**

51. Official printed or written statement that may be used as a proof or evidence of certain facts. **Certificate**

52. Ornamental hanging light with branches for several bulbs or candles. **Chandelier**

53. Fatty substance found in animal fluids and tissue thought to-cause hardening of arteries. **Cholesterol**

54. Regular pattern of weather conditions such as temperature, rainfall, winds etc. of a particular region. **Climate**

55. Social and economic system in which there is no private ownership and the means of production belong to all members of society. **Communism**

56. Fellow member of a trade union or of a socialist or communist political party etc. **Comrade**

57. Person who has done something wrong. **Culprit**

58. Person who believes that people do not do things for good, sincere or noble reasons but only for their own advantage. **Cynic**

59. Facts or information used in deciding or discussing something. **Data**

60. Academic title given by a university or college to one who has passed an examination, written a thesis etc. **Degree**

61. System of government by the people through the representatives they elect. **Democracy**

62. Ruler with unlimited powers especially a cruel and oppressive one. **Despot**

63. Person in the diplomatic service e.g. ambassador etc.

OR

A person clever at dealing with people. **Diplomat**

64. Illness of the body, of the mind or of plants caused by infection or internal disorder. **Disease**

65. Legal ending of a marriage, separation. **Divorce**

66. Weekly payment made by the state to unemployed people. **Dole**

67. Model of the human figure used for displaying or fitting clothes. **Dummy**

68. Device for converting steam-power, water-power etc. into electricity. **Dynamo**

69. Blocking of the light of the sun or moon due to specific position of the sun, the moon and the earth. **Eclipse**

70. Scientific study of the relation of plants and living creatures to each other and to their surroundings. **Ecology**

71. Science or principles of the production, distribution and consumption of goods especially with reference to cost. **Economics**

72. Person showing or holding a belief in equal rights, benefits and opportunities for everybody. **Egalitarian**

73. Minute particle of matter with negative electric charge found in all atoms. **Electron**

74. Poem or song expressing sorrow for the dead. **Elegy**

75. Art or style of speaking clearly and effectively especially in public. **Elocution**

76. Quickly spreading disease among many people in the same place for a time. **Epidemic**

77. Information that proves something or gives a reason for believing something. **Evidence**

78. Collection of things such as works of art, industrial or commercial goods etc. shown publicly for advertisement. **Exhibition**

79. Departure of many people at one time. **Exodus**

80. Introduced from another country; not native. **Exotic**

81. Person with special knowledge, skill, or training in a particular field. **Expert**

82. Spoken or done without previous thought or preparation. **Extempore**

83. Person more interested in what is happening around him than in his own thoughts and emotions. **Extrovert**

84. Department or group of departments in a university etc. **Faculty**

85. Extreme right-wing dictatorial political system or views (as originally seen in Italy between 1922 and 1943) **Fascism**

86. Related to government money or public money. **Fiscal**

87. All the plants of a particular area or period of time. **Flora**

88. Remains of a pre-historic animal or plant preserved by being buried in earth and now hardened like rock. **Fossil**

89. Border between two countries. **Frontier**

90. Type of sugar found in fruit-juice, honey etc. **Fructose**

91. Ceremony of burning or burying dead body. **Funeral**

92. Substance that kills fungus. **Fungicide**

93. Any of various types of plant without leaves, flowers or croon colouring matter. **Fungus**

94. Tube or pipe that is wide at the top and narrow at the bottom used for pouring liquids, powders etc. into small opening. **Funnel**

95. Enclosed space or chamber for heating metal glass etc. to a very high temperature. **Furnace**

96. Scientific study of the ways in which characteristics are passed from parents to their off springs. **Genetics**

97. Person who has exceptionally great mental or creative ability. **Genius**

98. Substance used for killing germs. **Germicide**

99. Person who holds esp. the first or bachelor's degree from a university. **Graduate**

100. Soft black substance used in making lead pencils, in lubrication and for slowing down neutrons in atomic reactors. **Graphite**

101. Any thick semi-solid oily substance used for lubrication. **Grease**

102. Room in a theatre, T. V. studio etc. where the performers can relax or do their make-up. **Green-room**

103. Protective screen of metal bars or wires. **Grill**

104. Shopkeeper who sells food in packets, fins, or bottles and general small household goods. **Grocer**

105. Person who shows others the way esp. a person employed to point-out interesting sights on a journey or places of tourist. **Guide**

106. Exercises performed to develop the muscles or fitness or to demonstrate agility. **Gymnastics**

107. Scientific study and treatment of diseases and disorders of the female reproductive system. **Gynecology**

108. Member of a wandering group of people who live in caravans. **Gypsy**
109. Substance carrying oxygen in the red blood-cells of vertebrates. **Haemoglobin**
110. Person with the legal rights to receive property when the owner dies. **Heir**
111. Animal that feeds on plants. **Herbivore**
112. Person who has withdrawn from society and lives completely alone. **Hermit**
113. Geometric figure with six sides and angles. **Hexagon**
114. Large scale destruction of life and property especially due to fire. **Holocaust**
115. Things said or done to show great-respect to a person or his qualities. **Homage**
116. The line at which the earth and sky appear to meet. **Horizon**
117. The scientific study of art and science of growing flowers, fruit and vegetables. **Horticulture**
118. Friendly and generous reception and entertainment of guests or strangers in one's own home. **Hospitality**
119. Person/s held as captive and threatened to be killed or harm to be done unless certain demands are met. **Hostage**
120. Animal or plant that has parents of different species or varieties. **Hybrid**
121. Study and practice of cleanliness as way of maintaining good health and preventing disease. **Hygiene**
122. Huge mass of ice floating in the sea. **Iceberg**
123. Phrase or sentence whose meaning is not clear from the meaning of its individual words in isolation but must be learnt as a whole unit. **Idiom**
124. Small dome-shaped house built by Eskimos from blocks of hard snow as a temporary shelter. **Igloo**
125. Not following accepted standards of morality. **Immoral**
126. That cannot be harmed by a disease or illness. **Immune**
127. List of names or topics referred to in a book etc. usually arranged at the end in alphabetical order. **Index**
128. Type of small animal having six legs, no backbone and a body divided into three parts viz. head, thorax and abdomen. **Insect**
129. Guarantee of compensation for loss, damage, sickness, death etc. in return of regular payment of premium. **Insurance**
130. An organization through which national police forces can cooperate with each other. **Interpol**
131. Meeting at which applicant is asked questions to find out if he is suitable for what he has applied. **Interview**
132. Enter a country or territory with armed forces in order to attack damage or occupy it. **Invade**
133. Supply water to land or crops by means of streams, reservoirs channels, pipes etc. **Irrigate**
134. Piece of land surrounded by water on all sides. **Island**
135. Creamy-white, bone-like substance forming the tusks of elephants, walruses etc. **Ivory**
136. Device for raising heavy weights off the ground esp. vehicles, motor cars etc. **Jack**
137. Hard, green stone from which ornaments are carved. **Jade**
138. Sweet substance made by boiling fruit with sugar until it is thick, preserved in jars. **Jam**

139. Technical or specialized words used by a particular group of people and difficult for others to understand. **Jargon**

140. Short, close-fitting jacket without sleeves worn by men or women. **Jerkin**

141. Aircraft powered by a jet engine. **Jet**

142. Newspaper or periodical especially one that is serious and deals with specialized subject. **Journal**

143. Celebration of a special anniversary of an event. **Jubilee**

144. Japanese system of unarmed combat in which arms, feet etc. are used as weapons. **Karate**

145. Metal instrument shaped to move the bolt of a lock. **Key**

146. Either of a pair of organs in the body that remove waste products from the blood and produce urine. **Kidney**

147. Of or produced by movement. **Kinetic**

148. Clothing and personal equipments of a soldier or of a traveller. **Kit**

149. Room or building in which food/meals are cooked or prepared. **Kitchen**

150. Room or building used especially for scientific research, experiments, testing etc. **Laboratory**

151. Existing but not active, developed or visible. **Latent**

152. Machine that shapes pieces of wood, metal etc. by holding and turning them against a fixed cutting tool. **Lathe**

153. Person who is trained and qualified in legal matters esp. solicitor. **Lawyer**

154. Story handed down from the past especially one that may not be true. **Legend**

155. Type of plants like pea, bean etc. which has its seeds in pods. **Legume**

156. Communication and co-operation between units of an organization. **Liaison**

157. False, written or printed statement that damages somebody's reputation or status. **Libel**

158. Freedom from captivity, slavery or oppressive control. **Liberty**

159. Room or building where the books for reading or borrowing are kept. **Library**

160. Liquid medicine or cosmetic for use on skin. **Lotion**

161. Oily or greasy substance that is put on or in machinery so that it moves easily. **Lubricant**

162. Meal taken in the middle of the day. **Lunch**

163. Person who writes the words of esp. popular songs. **Lyricist**

164. Apparatus with several moving parts for performing particular task and driven by electricity, steam, gas etc. or human power. **Machine**

165. Paper or periodical published every week or month with articles, stories etc. by various authors. **Magazine**

166. Done with or controlled by the hands. **Manual**

167. Animal dung or other material spread over or mixed with soil to make it fertile. **Manure**

168. Head of the council of a city or borough, usually elected yearly, who enjoys the status of the ' first citizen ' of the town. **Mayor**

169. A monument, ceremony etc. that reminds people of an event or person. **Memorial**

170. A heavy silver-coloured metal (chemical element) found in liquid form used in scientific instrument like thermometer, Barometer etc. **Mercury**

171. Science of the properties of metals, their uses, methods of obtaining them etc. **Metallurgy**

172. Scientific study of the earth's atmosphere and its changes used especially for forecasting weather. **Metrology**
173. A scientific instrument useful for making very small object appear larger. **Microscope**
174. Person who moves from one place to another to live or work. **Migrant**
175. Person in trade union or politics using force or strong pressure or supporting their use to achieve desired aims. **Militant**
176. Place where coins are made usually under state authority. **Mint**
177. Sole right to supply or trade in some commodity or service. **Monopoly**
178. Scientific study of the form and structure of animals and plants. **Morphology**
179. Person born in a place country etc. and associated with that place by birth. **Native**
180. Feeling of sickness or disgust for something. **Naseua**
181. Ornament of pearls, beads etc. worn round the neck. **Necklace**
182. Not supporting or helping either side in a dispute, contest, war etc. **Neutral**
183. Poisonous oily substance found in tobacco. **Nicotine**
184. Member of a tribe that wanders from place to place looking for pasture for its animals and having no fixed home. **Nomadic**
185. Person with official authority to witness the signing of legal documents and perform certain other legal functions. **Notary**
186. Book-length story in prose about either imaginary or historical characters. **Novel**
187. Central part of a living cell. **Nucleus**
188. Place where young plants and trees are grown for transplanting later and use for sale. **Nursery**
189. Substance serving as or providing nourishment especially for plants or animals. **Nutrient**
190. Fertile place with water and trees in a desert. **Oasis**
191. Thing in the way that either stops progress or makes it difficult. **Obstacle**
192. A long poem expressing noble feelings and often written to a person or thing. **Ode**
193. Smooth greasy paste for rubbing on the skin to heal injuries or roughness or as a cosmetic. **Ointment**
194. Suggesting that something bad is about to happen. **Ominous**
195. Path followed by a planet, star, moon etc. round another body. **Orbit**
196. A large group of people playing various musical instrument together. **Orchestra**
197. Of or from the countries of the East. **Oriental**
198. Scientific study of birds. **Ornithology**
199. Person whose parents are dead. **Orphan**
200. Branch of surgery that deals with correction of bone deformities and diseases. **Orthopaedics**
201. Instrument used to draw an exact copy of a plan, map etc. on any scale. **Pantograph**
202. Loss of control of a part of the body caused due to a disease of or injury to the nerves. **Paralysis**
203. Official document issued by the government of a country to its citizen to travel abroad under its protection. **Passport**
204. Person who gives money or other support to a person, cause, activity etc. **Patron**
205. Chemical substance used to kill insects. **Pesticide**
206. Deadly infectious disease that spreads quickly through large numbers of people. **Pestilence**

207. Implement with a curved blade used for digging furrows in the soil before sowing seeds. **Plough**

208. Artificial fabric used for making clothes etc. **Polyester**

209. Custom of having more than one wife at a time. **Polygamy**

210. Place where ships load and unload cargo. **Port**

211. Liked, admired or enjoyed by many people. **Popular**

212. Hens, ducks, geese, turkeys etc. kept for eating or for their eggs. **Poultry**

213. Amount or instalment regularly paid for an insurance policy. **Premium**

214. Study of the mind and how it functions. **Psychology**

215. Place where stones, slate etc. is extracted from the ground. **Quarry**

216. Bitter liquid made from the bark of a tree and used in drinks or as a medicine against malaria, fever etc. **Quinine**

217. Competition especially on T. V. or radio in which people try to answer questions to test their knowledge. **Quiz**

218. Repeat in speech or writing words previously said or written by another person. **Quote**

219. Number obtained when one number is divided by another. **Quotient**

220. Contest of speed between runners, horses, cars, vehicles etc. to see which reaches the specific place first. **Race**

221. Large gathering of people with a common purpose. **Rally**

222. Fixed quantity of food-grains etc. sold especially in times of shortage. **Ration**

223. Person who fights against or refuses to serve the established government. **Rebel**

224. Best performance or highest or lowest level ever reached especially in sports. **Record**

225. Exact copy made by an artist of one of his own pictures. **Replica**

226. Class of cold-blooded, egg, laying animals like lizards, crocodiles, snakes etc. **Reptiles**

227. Public place where snacks, meals, beverages etc. can be bought and eaten. **Restaurant**

228. Deliberate punishment or injury inflicted in return for what one has suffered. **Revenge**

229. Rise in rebellion against authority. **Revolt**

230. Overthrow of a system of government especially by force. **Revolution**

231. Wild or violent disturbance by a crowd of people. **Riot**

232. Person or thing competing with another. **Rival**

233. Part of a plant that keeps it firmly in the soil and absorbs water and food from it. **Root**

234. Information spread by being talked about but not certainly true. **Rumour**

235. Connected with or dedicated to God or religion. **Sacred**

236. Strong lockable box, cabinet etc. for storing valuables. **Safe**

237. Fixed, regular payment to employees doing other than manual or mechanical work. **Salary**

238. Liquid produced in the mouth that helps one chew and digest food. **Saliva**

239. An instrument with a long, sharp-toothed edge operated manually or mechanically for cutting wood, stone, metal etc. **Saw**

240. A frame of metal tubes or bamboo and wooden planks put up next to building so that builders, painters etc. can work on it. **Scaffold**

241. Act, behaviour etc. that causes public feelings of outrage or indignation. **Scandal**

242. Either of the two divisions of the academic year in a college or a university. **Semester**

243. Caused by or causing infection with harmful bacteria. **Septic**

244. Outer covering of eggs, of nut-kernels of some seeds or fruits and of animals like oysters, crabs, snails, tortoise etc. **Shell**

245. Method of writing rapidly using special quickly-written symbols. **Shorthand**

246. Short-handled tool with a curved blade used for cutting grass, corn etc. **Sickle**

247. Scientific study of the nature and development of society and social behaviour. **Sociology**

248. Tool for gripping and turning nuts on screws, bolts etc. **Spanner**

249. Figure of a person, an animal etc. in wood, stone, bronze, usually life-size or larger. **Statue**

250. Work-room of a painter, sculptor, photographer or a place where cinema films are acted and photographed. **Studio**

251. Doctor who performs surgical operations. **Surgeon**

252. Combining of separate parts elements etc. to form a complex whole. **Synthesis**

253. Armoured fighting vehicle with guns which moves on Caterpillar tracks. **Tank**

254. Sum of money to be paid by people, businessmen, workers etc. to a government for public purposes. **Tax**

255. Optical instrument shaped like a tube, with lenses to make distant object appear larger and nearer. **Telescope**

256. Formal offer to supply goods or carry out work at a stated price. **Tender**

257. Instrument for measuring temperature. **Thermometer**

258. Medicine that gives strength or energy, taken after illness or when tired. **Tonic**

259. Powerful motor vehicle used for pulling farm machinery or other heavy equipments. **Tractor**

260. Passing of beliefs or customs from one generation to the next esp. without writing. **Tradition**

261. Branch of mathematics dealing with the relationship between the sides and angles of triangle. **Trigonometry**

262. Underground passage for a road, or railway through a hill or under a river or the sea. **Tunnel**

263. Person appointed to see that the rules are observed during play etc. and settle dispute. **Umpire**

264. All existing things including the earth and its creatures and all the stars, planets etc. in the space. **Universe**

265. Institution that teaches and examines students, awards degrees, provides facilities for academic research. **University**

266. Awkward to move or control because of its shape, size or weight etc. **Unwieldy**

267. White soluble crystalline compound contained especially in the urine of mammals. **Urea**

268. Imaginary place or state of things in which everything is perfect. **Utopia**

269. Substance that is injected into the blood-stream and protects the body from disease.

	Vaccine

270. Space that is completely empty of all matter or gas/gases.	**Vacuum**

271. Wanderer or vagrant, especially an idle or dishonest person.	**Vagabond**

272. Conveyance such as car, lorry or cart etc. used for transporting goods or passengers on land.	**Vehicle**

273. Place where people agree to meet especially for a sports contest or match.	**Venue**

274. At a right angle to another line or plane or to the earth's surface.	**Vertical**

275. Metal tool with a pair of jaws that hold a thing securely tight while work is done on it.

	Vice

276. One who is alert and watchful against any possible danger, trouble in the offing.

	Vigilant

277. Person guilty or capable of great wickedness.	**Villain**

278. Using, showing or caused by strong physical force.	**Violent**

279. Mountain or hill with an opening or openings through which lava, cinders, gases come up from below the surface of the earth, may come up after intervals or cease to come up.

	Volcano

280. Document showing that money has been paid for goods, services etc. received.	**Voucher**

281. Regular payment made or received for work or services.	**Wages**

282. Four wheeled vehicle driven by horses or oxens, or open railway truck or trolly used for carrying heavy loads.	**Wagon**

283. Fighting between nations or groups within a nation using military force.	**War**

284. A small instrument showing the time, worn on wrist.	**Watch**

285. Thing designed or used for causing physical harm.	**Weapon**

286. Condition of the atmosphere at a certain place and time with reference to temperature, rain, sunshine, wind etc.	**Weather**

287. Disc or circular frame that turns or an axle as on vehicles or as a part of machine.	**Wheel**

288. Place in a river or the sea where there are whirling currents.	**Whirlpool**

289. Woman whose husband has died and who has not married again.	**Widow**

290. Mill worked by the action of wind on long projecting arms that turn on central shaft.

	Windmill

291. Process of producing photocopies without the use of wet materials.	**Xerox**

292. Type of short-wave electromagnetic radiation that can penetrate solids.	**X-ray**

293. Time taken by the earth to make one orbit round the sun $\left(\text{about } 365\frac{1}{4} \text{ days}\right)$	**Year**

294. Fungous substance used in making of beer and wine or to make bread.	**Yeast**

295. Round yellow part in the middle of the white of an egg.	**Yolk**

296. Point in the heavens directly above an observer. **Zenith**

297. Turning right and left alternatively at sharp angles. **Zigzag**

298. Imaginary band of the sky containing the positions of the sun, the moon and the main planets, divided into 12 equal parts. **Zodiac**

299. Area, band or stripe that is different from its surroundings or area or region with a particular feature or use. **Zone**

300. Place like garden, park etc. where living, especially wild, animals are kept for exhibition, study and breeding. **Zoo**

301. Scientific study of the structure, form and distribution of animals. **Zoology**

Part – V

SOFT SKILL DEVELOPMENT

(1) Soft Skills

(2) Speaking Skills

(3) Introduction to Group Discussion

(4) Process of Group Discussion

(5) Leadership Skill

(6) Instant Public Speaking

(1) SOFT SKILLS

Introduction:

'Soft skills' is a sociological term which refers to the cluster of personality traits, social graces, ability with language, personal habits etc. In other words, a set of skills that influence how we interact with each other are known as 'Soft Skills'. It includes abilities such as effective communication, creativity, analytical thinking, diplomacy, flexibility, change-readiness and problem solving, leadership, team building and speaking and listening skills. Personal management skills such as attitudes and behaviours that drive ones potential for growth and teamwork skills are also comprised in soft skills.

Meaning:

Soft skills include all facets of general skills that cover the intellectual aspect and non-academic skills of an individual. These skills help an individual in developing better interaction, performance and career prospects in an organisation.

Definitions:

1. *"Soft skills are personal management skills such as attitudes and behaviour that drives ones potential for growth and team work skills".*

2. *"Soft skills are the kind of skills needed to perform jobs where job requirements are defined in terms of expected outcomes."*

3. *"Soft skill is a set of skill that influences how we interact with each other. It includes such abilities as effective communication, creativity, analytical thinking, diplomacy, flexibility, change-readiness and problem solving, leadership, team building and listening skills."*

(5.1)

Importance of Soft Skills:

1. **Improves communication skills:** Communication is a medium to express ones thoughts, ideas, feelings, emotions, etc. The ability to communicate is highly essentially at every work place. Communication skill can be made more effective with the aid of developing soft skill in an individual. With the help of soft skills, an individual can deliver his idea clearly and confidently either orally or in writing. It also enhances usage of technology during presentations. Soft skills facilitate people from varied background to communicate in an effective way.

2. **Develops overall personality:** Soft skill is an aid to develop an overall personality of an individual in order to facilitate him for delivering his best performance at the work place. It also inculcates positive approach towards work among employees. It boosts the morale of the employees and makes them understand the need of the organisation professionally in terms of economic crisis, environmental, social and cultural aspects.

3. **Team Building:** Every individual working in an organisation is unique. Some employees may be comfortable working within a group, while others may prefer to work alone. Soft skill helps an individual to overcome these differences and build a good rapport and work effectively with others as a team. Moreover, it improves the ability of an individual to recognise and respect other's attitude, behaviour and beliefs.

4. **Problem solving ability:** Soft skill enhances an individual's ability to identify and analyse problems in critical situations and to come up with acceptable appraisal for the given problem. In addition to these, Soft skill develops an aptitude among individuals to find novel ideas and look for alternative feasible solution. It helps to assimilate and accommodate oneself to the diverse working environment.

5. **Customer service skills:** Customer service skill is one of the most essential skills an entrepreneur or business owner looks for while hiring important folks who will take care of their customer. Today's world is more customer centric which demands from an organisation to deliver customised services as per their expectations. In turn, to meet these expectations of their customers, the organisation concentrates on training their employees to improve their customer service skills such as developing patience, attentiveness, clear communication skills, knowledge of the product, time management skills, persuasion skills, willingness to learn, etc.

6. **Entrepreneurship skills:** Soft skills play an important role in the development of the employees as well as for the entrepreneur. Entrepreneur is the person who binds the employees with the organisation for the achievement of organisational goals. The entrepreneur should possess the soft skill of identifying, exploring, exploiting and building business opportunities and remain competitive in the critical business environment. Thus, soft skills are equally important for the entrepreneur to possess.

7. **Leadership skills:** The ability to lead effectively is based on a number of key skills. These skills are highly sought after by the employers as they involve dealing with people in such a way as to motivate, enthuse and build respect. Soft skill enables an individual to understand and take turn as a leader and provides him with capability to lead and supervise a team.

Elements of Soft Skill

Grooming Manners:

Grooming is the process of making yourself attractive and presentable. In simple words grooming means the things which you do to make yourself and your appearance tidy and pleasant. Personal Grooming is the term for how people take care of their body and appearance. Habits that are considered in personal grooming include dressing, make up, taking care of one's teeth and skin, etc. Appearance, clothes and manners do not make the man; but, when he is made, they greatly improve his appearance whether this is real or imaginary; the most important fact is that your appearance influences the opinions of everyone around you. Your professionalism, intelligence and the trust people form in you depends on your appearance.

Importance of Grooming

- In the business atmosphere it is extremely important to have knowledge of grooming skills because you deal with people of different cultures on a regular basis and should always aim to make the best first impressions, lasting impressions or you risk losing potential opportunities for yourself.

- A person with bad personal grooming habits can be spotted from anywhere. Therefore, personal grooming is highly essential.

- A clean and neat appearance inspires confidence. Grooming is important if you want to feel confident and project a positive self-image of yourself. For a good impression, grooming is must.

- Grooming is an external appearance, which is the window of your personality to the world. External appearance is very important as it portray the first impression to others about your personality.

What is Etiquette?

Business etiquette is a set of rules that govern the way people interact with one another in business, with customers, suppliers, with inside or outside bodies. It is all about conveying the right image and behaving in an appropriate way. Etiquette refers to behaving in a socially responsible way.

Etiquette refers to guidelines which control the way a responsible individual should behave in the society

Need for Etiquette

- Etiquette makes you a cultured individual who leaves his mark wherever he goes.

- Etiquette teaches you the way to talk, walk and most importantly behave in the society.

- Etiquette is essential for an everlasting first impression. The way you interact with your superiors, parents, fellow workers, friends speak a lot about your personality and up-bringing.

- Etiquette enables the individuals to earn respect and appreciation in the society. No one would feel like talking to a person who does not know how to speak or behave in the society. Etiquette inculcates a feeling of trust and loyalty in the individuals. One becomes more responsible and mature. Etiquette helps individuals to value relationships.

Types of Etiquette

1. **Social Etiquette:** Social etiquette is important for an individual as it teaches him how to behave in the society.

2. **Corporate Etiquette:** Corporate etiquette refers to how an employee should behave while he is at work. Each one needs to maintain the dignity and the decorum of the organisation.

3. **Meeting Etiquette:** Meeting etiquette refers to a manner one need to adopt when he is attending any meeting, seminar, presentation and so on. Listen to what the other person has to say. Never enter a meeting room without a notepad and pen. It is important to jot down important points for future reference.

4. **Telephone Etiquette:** It is essential to learn how one should interact with the other person over the phone. Telephone etiquette refers to the way an individual should speak on the phone. Never put the other person on long holds. Make sure you greet the other person. Take care of your pitch and tone.

5. **Business Etiquette:** Business etiquette includes ways to conduct a certain business. Don't ever cheat customers. It is simply unethical.

Some basic etiquettes to be followed in Business Dealings

1. **Introduction:** Always introduce your staff to others whenever the opportunity arises, unless you know that they are already acquainted. It makes people feel appreciated, regardless of their rank or position.

2. **Being polite:** Being polite is the basic form of courtesy highly essential in a casual professional atmosphere. It is always nice to send a handwritten thank you note as it gives an emotional touch rather than a formal email conveying thank you.

3. **Not to be over aggressive:** People by nature are always eager to offer their opinions or often interrupt others when they speak. It is rude and shows disregard for the opinions of others. Remember, be assertive, not aggressive.

4. **Proper language:** In business dealings, always one should be careful to choose words wisely. Rude, offensive or slang language is undesirable.

5. **Don't chitchat:** Gossiping at business place can lead to adverse problems for self and others. Gossips are harmful for the organisation. One should try to avoid talking about someone who is not present.

6. **Don't spy:** Everyone is entitled to private conversations, in person or over the phone. The same goes for e-mail; don't stand over someone's shoulder and read their e-mails.

7. **Recognise others:** At the workplace when an employee approaches you, acknowledge him or her. If you are in the middle of some important work, it is all right to ask them to wait a minute while you finish your work. Being busy is not an excuse to ignore others.

8. **Punctual:** Being punctual shows others that you value their time. One should always try to avoid getting late for the work entrusted to them. Punctuality creates disciplines in the organisation.

9. **Avoid phone calls in Meetings:** When you are in a meeting, do not attend a phone call, which distracts your focus on the meeting. It also makes meetings last longer because the partaker of the meeting keep losing focus.

10. **Paying attention when others speak:** Effort should be made to truly listen to what others say. One should put an effort and patiently wait for the other person to finish their talk and then one can move on to the next thing. Take the time to respond to others queries and show an interest in the other person's thoughts.

(2) SPEAKING SKILLS

Meaning of Effective Speaking:

The sound of a voice and the content of speech can provide clues to an individual's emotional state and a dialect can indicate their geographic roots. The voice is unique to the person to whom it belongs. For instance, if self-esteem is low, it may be reflected by hesitancy in the voice, a shy person may have a quiet voice, but someone who is confident in themselves will be more likely to have command of their voice and clarity of speech.

Effective speaking has nothing to do with the outdated concept of 'elocution' where everyone was encouraged to speak in the same 'correct' manner. Rather, effective speaking concerns being able to speak in a public context with confidence and clarity, whilst at the same time reflecting on your own personality.

Elements of Good Speaking:

It is important to follow certain principles to make sure that one is able to get the message across.

1. **Clear Pronunciation:** The first important prerequisite of effective oral communication is that the words should be pronounced clearly and correctly. Oral messages are often misunderstood because the speaker does not talk distinctly. Inability to use the jaws freely, to speak with limber tongue and limber lips and speaking slowly often makes for poor oral transmission. If a person tries to talk as fast as he thinks, his words will run to gather and get rammed into one another. So, when he intends asking, 'What did one have?' he will succeed only in saying, `wajuhave'?

2. **Brevity:** People take pleasure in talking, so oral communication tends to suffer from over-communication. But, if a speaker keeps on talking for long, his message will get lost in a sea of verbosity and distraction. It is important to keep the message as brief as possible without appearing abrupt and discourteous.

3. **Precision:** Precision can make oral communication very effective. Instead of saying, 'Total these invoices as early as possible'; it is preferable to specify the time and say, 'Could you kindly total these invoices and bring them back to me in half an hour.' 'Come to the office early tomorrow', is not as good as, 'Could you reach the office tomorrow by 8 o'clock, since all these letters have to be dispatched by the first mail.'

4. **Conviction:** A person communicating orally must have conviction in what he says. Lack of conviction causes lack of confidence and so he is not able to impress the receiver with the message. Conviction comes from sincerity of approach and careful thinking and

planning. Careful analysis and objective evaluation of the message while formulating it also promotes the speaker's conviction in it.

5. **Logical Sequence:** If the speaker has given a proper thought to his message, he will be able to arrange various ideas contained in it in their logical sequence. Jumbled ideas create confusion, while logically arranged ideas make the message forceful.

6. **Appropriate Word Choice:** Words have different meanings for different people. So, it is important to be careful in the choice of words. The speaker, while speaking something, knows what he means and so presumes that his listener also does so, which may be a wrong presumption. In oral communication, it is more important to use the terms familiar to the listener rather than the terms that are familiar to the speaker.

7. **Avoiding Hackneyed Phrases and Clichés:** Speakers, when they are groping for words, often make use of everyday phrases like 'what I mean', 'do you follow', `isn't it', 'I see' etc. Such words and phrases interrupt the flow of their speech and impede the quick grasping of meaning. They are used unconsciously, but the speaker should take deliberate pains to exclude them from their speech.

8. **Natural Voice:** Some speakers deliberately cultivate an effected style under the impression that it would make them look more sophisticated. Nothing is farther from the truth and nothing impresses so much as the natural way of speech. One of the manuals for office employees in firm says, "The most effective speech is that which is correct and at the same time natural and unaffected. Try to tone down an unusual accent and discard all affectations of speech. Try to cultivate a pleasing voice and speak clearly and distinctly."

Types of Speaking:

Employees have to speak in an organisation with top and lower level management along with their peers, which helps them to get the work done from others. Better speakers upgrade their career graph and good speaking is a short cut to success. There are different types of speaking in an organisation which are as follows:

1. **Conversation:** A conversation is communication between multiple people. It is a social skill that is not difficult for most individuals. Conversations are the ideal form of communication in some respects, since they allow people with different views on a topic to learn from each other. A speech, on the other hand, is an oral presentation by one person directed at a group.

    For successful conversation, the partners must achieve a workable balance of contributions. A successful conversation includes mutually interesting connections between the speakers or things that the speakers know. For this to happen, those engaging in conversation must find a topic on which they both can relate to in some sense. Those engaging in conversation naturally tend to relate the other speaker's statements to themselves. They may insert aspects of their lives into their replies, to relate to the other person's opinions or points of conversation.

The majority of conversations can be divided into four categories according to their subject content:

- **Conversations about subjective ideas**, which often serve to extend understanding and awareness.

- **Conversations about objective facts**, which may serve to consolidate a widely-held view.

- **Conversations about other people** which may be critical, competitive, or supportive.

- **Conversations about oneself**, which sometimes indicate attention-seeking behaviour.

2. **Oral Presentation:** A presentation is a speech on a serious topic; its purpose is to inform, to explain, and to persuade the audience or to present a point of view. It may introduce a product or explain a process or narrate an experience; it is delivered to a small, knowledgeable audience at a conference, a seminar or a business meeting. It is followed by questions from the audience.

3. **Speeches:** Speech is an act of speaking or act of expressing or describing thoughts, feelings, or perceptions by the articulation of words. Speech is something spoken and an utterance or vocal communication or conversation. It is also a talk or public address.

4. **Dialogue:** A dialogue is a conversation between two or more people. It is also a literary form in which two or more parties engage in a discussion. Some employees are introvert and are not interested in a dialogue with anybody. This will hamper the growth of an organisation. A dialogue is based on a situation or a circumstance in an organisation. One has to be a good listener in the process of a dialogue.

5. **Group Discussion:** Group Discussion as the term itself suggests, is a discussion by and among a group of people. The group have between 8 and 12 members who express their views, freely, frankly and in a friendly manner, on a topic of current and controversial nature. The topic is given on the appointed date and hence, the conversation that follows is a spontaneous chit-chat, not a pre-planned one. Here, within a time limit of 20 to 30 minutes, the abilities of the members of the group are measured in an unobtrusive manner by the examiner who does not actively participate in the discussion.

Presenting the Speech:

A well prepared speech can go waste if it is not presented well. The success of all the efforts put in by the speaker depends on how skillfully he presents it. At this time, he will need to decide on his method of presentation. There are three methods or ways of presenting a speech, presenting it extemporaneously, by reading it, or by memorizing it.

Ways of Delivering the Speech:

1. **Presentation by reading:** Usually, the inexperienced speakers use this method, as lack of confidence does not allow them to memorize even a part of the speech. Unfortunately, most of us do not read aloud well. We tend to read in a dull monotone voice, producing a most uninteresting and larkluster effect. We fumble over words, miss punctuation marks and make similar lapses. Many speakers overcome this problem with effort, and eliminate it.

2. **Extemporaneous presentation:** It is the most popular and effective method of presentation. Using this method, the speaker initially prepares his speech. Then he prepares notes and presents the speech from them. This allows him to have good eye contact, while he may feel confident, having the support of the notes with him.

3. **Memorized presentation:** It is the most difficult method of presentation. Probably, a few speakers actually memorize an entire speech. Memorized speech does have poor display of non-verbal cues. The fear of forgetting the speech in between, is a big hurdle and does not allow the speaker to be at ease. Instead, memorizing key parts and using notes to help through the presentation, is a better option.

Important Aspects related to Presentation of Speech

Other aspects relating to presentation of the speech, which the speaker should be aware of, are:

1. **Appearance and body actions:** The listeners hear the speakers words, they are looking at him. What they see, is part of the message, and it can have real effect on the success of his speech. The audience sees, the speaker as also what surrounds him. Thus, in his efforts to improve the effects of his oral presentations, the speaker should also pay attention to his appearance and bodily actions.

2. **The communication environment:** Much of what his audience sees is, all that surrounds him as he speaks. All of that tends to add to a general impression. This includes physical things like the stage, lighting and, background and his own experience as a listener will tell him, what is important.

3. **Personal appearance:** The personal appearance of the speaker is part of the message. The audience receives most of the non verbal cues from their making is necessary that he uses it appropriately. Specifically, he should dress appropriately for the audience and occasion.

4. **Posture:** Posture or body position is likely to be the most important thing, which the audience sees about him. Even if listeners cannot be close enough, to detect facial expressions and eye movements, they can see in general the structure and state of the body. The speaker probably thinks that one should tell him what good posture is? He may know it when he sees himself. He should keep his body erect without appearing stiff and uncomfortable. His deportment should be poised, alert, and communicative. He should do all this naturally. The greatest danger with posture is appearing artificial. People may become too artificial pretentions or unspontaneous by reading books on communication.

5. **Walking:** The way the speaker walks before his audience also makes an impression on his listeners. A strong and sure walk both gives an impression of confidence. Walking during the presentation can be both good and bad, depending on how the speaker does it through in public speech? We rarely find speakers walking.

6. **Use of voice:** Good and effective voice is an obvious requirement of good speaking. Like bodily movements, the voice should not hinder the listener's concentration of the message. Voices that cause such difficulties, generally fall into four areas of fault.

7. Avoid a few words or phrases:

1. Technical terms

2. Latin and French words

3. Cheap, hollow and slang terms

4. Socially unpleasant words

5. Repeating phrases like "you see", "you know", "I mean" and such like.

6. Difficult words

While actually delivering the speech, the speaker may make use of visual aids, notes or his own manuscript.

Developing Confidence and Overcoming Fear

Public speaking will not give any results, if the speaker feels scared to face the audience. The speaker observes some signs of discomfort like increase in heart rate, rise in blood pressure, rise in body temperature, shivering of legs and hands, fumbling for words and sweating of palms. These are signs of nervousness and a lack of self-confidence. The reviews should help him to pinpoint problem areas and also give him practical suggestions to overcome them.

How to Overcome Stage Fear?

Fear is the manifestation of our own mind. A feeling that one knows the subject matter better than anyone else and that he is in charge, infuses enough confidence in the speaker to overcome fear.

Overcoming stage fear:

1. **Pre-check the equipment:** The projector, screen, display board, etc., required during the speech, should be checked prior to the speech. If at the time of the speech, any of the equipment does not work, it leads to humiliation of the speaker and loss of impact on the audience.

2. **Know your subject well:** Prepare with the attitude that you should know the subject better than anyone else.

3. **Rehearse several times:** At least a few complete rehearsals help to memorize the subject matter and be confident. It also helps to improve the non verbal part of the speech.

4. **Maintain poise and enthusiasm:** The speaker should maintain his poise, body posture and should emanate a good level of enthusiasm from his body language. This will help him to be friendly towards the audience.

5. **Breathe deeply and slowly before speaking:** A controlled breath helps to control the heartbeat and thus remove signs of nervousness. It also calms down the mind, thus preparing the speaker to face the audience.

6. **Move during the speech:** Some movement during the speech holds the attention of the audience and helps the speaker to release his stress.

7. **Carry the notes:** The speaker's notes, clearly written, should be carried by the speaker, for reference, before and during the speech. This helps to overcome nervousness.

Qualities to be Possessed by a Speaker:

The speaker should posses the following qualities to be an efficient and eloquent speaker:

1. **Confidence:** Even the most confident speakers have the nervousness whenever they occupy a stage for public speech. A primary characteristic of effective oral reporting is confidence. This includes his confidence in himself and his audience, in him. Actually, the two are complementary to each other. For example, he can prepare his presentation diligently and practice it thoroughly. Such careful preliminary work will give him confidence in himself. Unfair and logical as it may be, certain style of dress and hair create strong images in people's minds. Thus, if he wants to communicate effectively, he should analyze the audience to whom he seeks to reach.

2. **Sincerity:** The speaker must be sincere for listeners, always appreciate sincerity in the speaker. The listeners will be quick to detect insincerity in the speaker. When they do so, they are likely to give little weight to what he says.

3. **Thoroughness:** The speaker must be thorough regarding the subject matter of public speech. Thoroughness in his presentation will ensure that his message is better received than one, which is a hurried coverage. Thorough coverage gives the impression that he has taken proper time and adequate care, and such an impression tends to make the message believable.

4. **Friendliness:** A speaker who projects an image of friendliness ha a significant advantage in communication. Like sincerity, friendliness is difficult to pretend. It must be honest if it is to be effective. But with most people who friendliness is an honest effort for there are people want to be friendly too.

Handling Questions

Following guidelines will help to answer questions effectively. The speaker must listen to the question, raised carefully:

1. He should be sure about his understanding of the question. He should have clarity before he attempts to answer it. If he is not sure what a question means, ask the questioner to clarify it.

2. If a listener is asking several questions at a time, answer them one by one. If the answer to his first question is very long, the speaker can ask the listener to repeat his or her other questions.

3. He should make his answers direct and understandable. He should limit his answer to the question, which has been asked. Proper, traditional language is to be used.

4. He should ensure that the listener understands his answer.

5. If the speaker doesn't know the answer to a question, he should clearly tell the audience so, and should not bluff the listeners.

6. He should treat people who ask question with respect, even if they don't treat the speaker that way.

(3) INTRODUCTION TO GROUP DISCUSSION

Meaning of Group Discussion:

Group discussion, as the term itself suggests, is a discussion by and among a group of people. The group has between 8 and 12 members who express their views, freely, frankly and in a friendly manner, on a topic of current and controversial nature. The topic is given on the appointed date, and, hence, the conversation that follows is a spontaneous chit-chat, not a pre-planned one. Here, within a time-limit of 20 to 30 minutes, the abilities of the members of the group are measured in an unobtrusive manner by the examiner who does not actively participate in the discussion.

Group discussion is mostly unstructured. That is, every single step is not planned in advance. Each candidate is not given a time limit for speaking. Similarly, the order of speaking, that is, who will speak first and who will speak last is not fixed in advance. The candidates have to decide how to conduct the group discussion. The selectors see how the group takes shape, and who contributes most to it. They also judge the knowledge of each candidate, time management, leadership quality, behaviour, etc.

Do's and Don'ts of Participating in Group Discussions:

Do's

1. Keep to the point.
2. Make original points and back them by substantial reasoning and not an unfounded opinion.
3. If some other member has already made the point you wanted to make, do not worry. Even here, you can either support or oppose
4. Listen to the other participants patiently, attentively and carefully
5. Whatever you say must be with a logical flow.
6. Make only accurate statements.
7. Modulate the volume, pitch and tone of your voice
8. Show flexibility in your views.
9. Be considerate to the feelings of others.
10. If possible, make an attempt at initiating the discussion. Not only this, you must talk as much as possible, effectively and sensibly, to impress others
11. If you have reasonable doubts, you can ask questions
12. Try to get your turn. Don't think someone else will offer you an opportunity to speak.
13. As soon as you find your argument has been refuted by someone else, you must try to justify your point of view by bringing in fresh ideas or arguments.
14. If more than one candidate starts speaking simultaneously. You can request for order and appeal for co-operation.
15. Be an active and interested participant.
16. Listen to the initial briefing examiner carefully.
17. While speaking, address to the whole group, looking each participant in the eye in turn.
18. Be attentive to the person who is speaking.
19. Talk with confidence and self-assurance as a mature person.
20. Sit straight and upfront. To emphasise your point, make use of hands and facial.

21. If the discussion is straying from the ejected course, take the initiative of bringing it back on track by giving a new directional point.

22. What you say must be audible and clear to the group.

23. Your voice must be crisp, energetic and forceful.

24. If the discussion is stuck on one point for too long, the quality will definitely suffer, so rectify it.

25. If you find that you have sufficient knowledge of the subject, try to speak first or then speak later.

26. If you have not got chance, then request for a hearing.

27. In situations where the group has to make a choice of the subject for discussion, you take the lead in determining the topic.

28. In case you find that you have no ideas or have little knowledge of the subject, just concentrate on what is being said by others.

29. Even if you realise that what you are speaking does not cut much ice with other candidates, do not give in.

30. Last, but not the least important, develop the habit of discussing different topics with your friends and the members of your family.

Don'ts

- Initiate the discussion if you do not have sufficient knowledge about the given topic.
- Over speak, intervene and snatch other's chance to speak.
- Argue and shout during the GD.
- Look at the evaluators or a particular group member.
- Talk irrelevant things and distract the discussion.
- Pose negative body gestures like touching the nose, leaning back on the chair, knocking the table with a pen etc.
- Mention erratic statistics.
- Display low self confidence with shaky voice and trembling hands.
- Try to dominate the discussion.
- Put others in an embarrassing situation by asking them to speak even if they don't want to.

Guidelines for Group Discussions:

1. Never fail to do homework. Study the topic, collect information and gather points of view for each item.

2. Keep an open mind. However well-prepared you may be, there is a possibility that you may be wrong. Be prepared to learn and correct any mistake in thinking or information.

3. Do not disturb other participants or yourself by talking on the side or by shuffling papers.

4. Have a sporting spirit. If your idea/suggestion is defeated in the discussion, be graceful and thank others for helping you to clarify your ideas.

5. Show interest in what others say. When someone makes a good point, show appreciation even if it demolishes your point.

6. Never personalise a difference of opinion. If it is necessary to disagree with something that is said, first re-state or summarise it and then explain why you disagree.

7. Speak up if you have something to say, especially on a topic on which you have knowledge. You must be willing to contribute and share. But keep your comments short and precise.

8. Do not be carried away by emotions. Problems cannot be solved by anger.

9. Be willing to examine the ideas presented.

10. Be a good listener. You will learn a great deal about the topic and about human behaviour if you listen.

(4) PROCESS OF GROUP DISCUSSION

'Group Discussion' is one type of formal meeting. It can be a very stimulating and useful activity in an organisation. It helps in understanding a situation, in exploring possibilities and in solving problems as it generates multiple viewpoints. It gives a sense of participation to all those who participate in it. Such meetings can be based on a topic or on a case study.

Group discussions are widely used in many organisations for decision making and problem solving. Organisational group discussions are: Brainstorming, Nominal Group Technique and Delphi Technique.

It is used as a tool for selecting candidates by observing their behaviour and abilities in taking part in it.

Characteristics of Group Discussion

- Evaluation Components
- Knowledge
- Communication skills
- Active Listening
- Clarity of Thought and Expression
- Apt Language
- Appropriateness of Body Language
- Group Behaviour
- Leadership Skills

Group Discussion for Selection Process

Normally a group consists of 8 to 10 students. The normal time duration for any GD is about 10 to 15 minutes. For a GD relating to any particular topic, 2-3 minutes thinking time may be given. For GDs relating to case studies, 15 minutes time should be given for studying the case. The valuation of the participants is normally done by experts (usually professors from business schools). All these experts possess vast experience and expertise in their respective fields.

Essential Requirements for participating Effectively in GDs

1. Give proper respect to the views expressed by other speakers.

2. Maintain courtesy while speaking.

3. Do not use very insulting, harsh and aggressive language.

4. Speak clearly and pleasantly to all the members of the group.

5.   Stick to the main point of the GD. Do not speak irrelevant matter in the discussion.

7.   Do not use loud and angry tone while speaking.

8.   Do not interrupt other members of the group while they are speaking.

9.   Learn to disagree politely.

10.   Use positive body language while speaking.

Note on GD for Candidate Selection

There are no fixed rules for a GD held for candidate selection. The participants are generally eager to be the first to speak. The first speaker should mention the topic and state the issues. He should not give any opinion at this point. Later, the person may offer clarifications. This should be in the form of a statement, and not a question. Care should be taken to address the entire group. The candidates should maintain calm even if there are opposing arguments. The candidate should acknowledge others' viewpoints.

Situations may arise, when one's point of view has already been expressed by someone else. The best approach would be to acknowledge the contribution of the other person and add to it and thus become the centre of attention.

If there is an interruption from someone else, the approach should be to request the person to allow you to complete your statement and offer the field to that person.

The group should move towards a consensus but due to the all round anxiety to make one's point, this may not happen at all. The idea is to exhibit some leadership qualities in steering the group while making one's contribution. A person with leadership qualities will look at the clock and when it is almost nearing closing time, may start summing up. If the group is too noisy, the facilitator may allot one minute to each candidate to sum up the discussion. This is an opportunity to put on one's best effort. Without criticising the group, one can sum up and give one's own views. The candidates should not stray from the topic.

The candidates are evaluated on how they speak. Fluency, meaningful contribution, depth of knowledge on various topical issues, and leadership qualities are what the selectors would be looking for. Whatever personal views one may have, it is important to know both sides of the argument. One must be able to defend one's viewpoints convincingly and therefore the need for acquiring wide knowledge. Candidates should select topics, prepare and talk in front of the mirror, family members and friends to understand where they need to strengthen their arguments or presentation skills. They can also record their speech to know where they falter, and this will indicate the areas for further improvement. The voice quality, body language etc., can also be assessed during this process. The practice sessions can be used for ensuring that the thought process is well organised. One way to practice this is to write down the points and keep it in front of you. By periodically looking at it, you can arrange your thoughts mentally.

(5) LEADERSHIP SKILL

What Are Leadership Skills?

When we talk about leadership skills, what exactly do we mean? Leadership skills are the tools, behaviors, and capabilities that a person needs in order to be successful at motivating and directing others. Yet true leadership skills involve something more; the ability to help people grow in their own abilities. It can be said that the most successful leaders are those that drive others to achieve their own success.

A Born Leader?

You've certainly heard the phrase. Who do you think of when you hear it? Martin Luther King Jr., Mahatma Gandhi, other world-famous leaders in history? Or perhaps there are leaders in your own life that have had a positive impact on you. What skills did all of these people have that made them effective leaders? Here are a few, but there are certainly others:

- Is Committed to a Vision or Mission
- Understands His or Her Role
- Demonstrates Integrity
- Sets an Example
- Understands How to Motivate the Behavior of Others
- Communicates Effectively
- Is Willing to Take Risks
- Is Adept at Problem-Solving

You don't have to be born with leadership skills. They can be learned.

Whereas many leaders may be so committed to a vision that they naturally find ways to pull others along with them, most of us cannot claim to have been born with that level of leadership ability. We certainly may have grown over time and learned many effective skills by experience.

There is good news for anyone who doesn't consider themselves a born leader or who has specific areas of leadership skills that need work; leadership skills can be learned. All that is required is an open mind, patience with yourself as you learn these skills, and the commitment to put what you learn into action.

2. Three Traits Every Successful Leader Must Have

Introduction:

Without exception, there are three traits that every leader must have in order to be successful. You can attempt to lead without them, but at least one of four things will eventually happen if you do:

- You will be so miserable that you will burnout.
- Your team will fail in completing their work.
- Your team members will leave.
- Your team will lose respect for you.

So what are these three required traits? First is the desire to lead. Without it, you will never be comfortable in the leader role. You will struggle every day with the basics, and your team members will sense it in everything you do. If you don't burn out first, you'll find that work suffers and your team is frustrated because they can't do their work without you doing yours. They may eventually leave — if you don't first.

The second trait of successful leaders is commitment to the mission and vision of the organization where they work. Imagine trying to convince others to give their best in order to accomplish something they don't believe in. That's difficult. But trying to convert them to believing in the mission and vision of an organization when you don't believe it yourself? That's simply impossible.

The final trait that every successful leader must have is integrity. Integrity in this sense has a simple meaning; doing what you say you will do and behaving the way that you expect your team to behave. At first glance, that may sound simple enough. But if you can truly master integrity, you will find that it changes whole teams and even whole organizations for the better.

1. The Desire to Lead

As with any job, resisting the work of leading will make it difficult to be effective — and impossible to find fulfillment or enjoyment in what you do. Without the desire to lead, you will not be willing to do the work that it takes to become the leader of the team. You won't put in the effort to acquire the skills you need to motivate others or to handle conflict. Instead, you'll stick with the comfortable patterns of behavior you've already developed, regardless of whether or not it helps you to lead.

Are you sure that you want to be a leader? If you are, then you are already a step ahead. But if you aren't certain that you want to be someone who is followed rather than someone who follows, you need to consider whether or not a leadership role is right for you.

There are a number of characteristics and feelings that can help you determine your level of desire to lead. Given below is a series of statements which describe the characteristics which generally indicate your motivation to be a leader. Don't worry if you don't meet all of these criteria. Just use them to gauge where you are now in your leadership development.

- I enjoy it when others seek my ideas or opinions.
- I don't mind asking team members challenging questions when working on a project.
- I like supporting others on my team and can do so in both good and bad times.
- I am comfortable putting the team interest before my own interest.
- When I am working with a group, I facilitate a strong team spirit.
- I am comfortable letting others take my ideas and put them into action.
- I like playing the role of coach to help others improve their skills.
- I prefer to resolve personal conflicts on a team rather than let them continue.
- I look for opportunities to celebrate other people's success.
- I can have a productive discussion even when others don't agree with me.
- When the team has a problem, I consider it my problem too.
- I like to generate ideas and share them with the group.

It is possible to have these and similar statements apply to you in some companies and not with others. If you don't like the work that you do, chances are that you are not going to be inspired to lead others to do it either. In that case, you won't reach your potential as a leader until you are working for an organization that you can believe in. This brings us to the second trait that successful leaders all possess.

2. Commitment to the Mission and Vision of the Organization

The first leaders of any organization were those that first created it. They had a mission and a vision about what the company would do, who it would serve, and what changes it would make to the industry or sector they were entering. Those leaders had to take risks like borrowing money or leaving the job they were in at the time in order to start the business. There were likely personal sacrifices, long nights and weeks of work, and times of significant stress before the company could be considered a success.

Those first leaders then had to hire others who could share the vision and believe in the mission of the company. The leaders would coach these new hires, helping them learn how to make decisions which would move the team or organization towards the company's goals. This trend would hopefully continue as each layer of administration was added to the organization. In an ideal world, every employee would perform as if fulfilling the mission and vision of the organization were his or her personal goal.

Of course, we don't live in an ideal world. The more layers of administration or bureaucracy there are between the visionaries at the head of an organization and the front-line employees who deliver the actual services or product, the more difficult it is to see the mission and vision translated to the employees.

At this point, there is one question you need to ask yourself. Do you know what the mission and vision of your organization are? If you responded by naming what you do, that's not the same thing. For example, a telecommunications company employee might answer, 'we sell telecommunication products.' But is that the mission of the organization? The mission could be something like 'we help people stay connected by providing the highest quality communication products and the best customer service in the industry.' Can you see the difference? One is what you do. The other is how you do it. If you know what you do, but not the way you are expected to do it, you cannot effectively lead others to assist in accomplishing the company's goals.

3. Integrity

The third trait that every leader must have in order to be successful is integrity. Integrity can be defined simply as being true to your word, being authentic in your actions and speech, and demonstrating the kind of behavior that you would like to see your employees have. Integrity, like leadership skills, is something that you have to practice. It takes effort to honor your word every time and to be the example you want from your employees even when you are under stress or simply have a personality conflict. But the benefits you can gain from developing integrity are enormous when compared to the damage you can do in the workplace if you lack it.

Think for a moment about characteristics of bosses you have had that you didn't like. What, specifically, were the attitudes, behaviors, or traits of that person that has you still thinking of them in a negative light? Probably you would list things like favoring certain employees, not coming through on promises he or she made, gossiping, taking credit for your work, or treating you disrespectfully. All of these issues can be traced to a lack of integrity.

What was the workplace environment like when you worked for someone who lacked integrity? Did you enjoy going to work? Did you feel inspired to take ownership of your projects and put forth the best effort that you could? Did you feel loyal to the company or believe that there were significant chances for your own personal growth and development? It's a safe bet that your answer to these questions is no. And it's just as safe a bet that as a leader, you could be creating the same kind of environment that you hated if you aren't practicing integrity in the workplace.

So how do you practice integrity? There are three key areas that you can concentrate on developing. As you read each description, ask yourself how you would feel if a leader you worked for did not possess these key characteristics.

(i) Sincerity

Also called authenticity, leaders with this facet of integrity:

- Do not put up a false front.
- Accept responsibility for their commitments and strive to meet them.
- Are honest about their own limitations.
- Accept responsibility for their mistakes.
- Tell the truth.

(ii) Consistency

Leaders demonstrate this facet of integrity by:

- Treating employees equally as much as possible.
- Following through on promises.
- Working as hard or harder than their employees.
- Having the same expectations or rules for themselves as for their employees

(iii) Substance

Substance refers to integrity becoming a part of who you are being in all your work relationships by:

- Keeping private employee information private.
- Not gossiping or complaining about team members to other team members.
- Doing what's best for the team and not just yourself.
- Giving credit where credit is due.
- Caring about the development of your employees.
- Making it a priority to maintain clear communication and resolve any conflicts.

If you have read this information and realized that you have not always acted with integrity in the workplace, you are certainly not alone. But going forward, you can now recognize that integrity can be built one action at a time. As you get more practiced at it, you will find that it becomes a habit. And once you start seeing the results that come from practicing integrity, you will want to keep going.

"The supreme quality for a leader is unquestionably integrity." **– Dwight D. Eisenhower**

1. **Leadership Qualities:**

   (i) **Taking initiative:** Primary among the qualities looked for, is the leadership quality. In GDs leadership qualities necessitates that the individuals have the capacity to take initiative during the course of the interaction. This could entail adopting strategies such as beginning the discussion, picking up the threads at a later stage, etc.

   (ii) **Ability to give direction:** It is not necessary to take the initiative if one is not familiar with the topic. The quality which comes subsequent to it, is the quality of possessing the ability to give direction to the entire discussion. It follows naturally that the interactant should have the power and ability to sum up all, that is being said in a manner which is conducive to the growth of the discussion. The essential attributes of a leader are, therefore, to give direction an individual would be viewed as a leader with the capacity of chalking out a strategy, filtering and assimilating ideas while leading and controlling the interactants.

(iii) Taking the group along: The ability to sum up discussion not only at the end but also in between, is a major quality of the interactant. It helps in preventing the group from straying away from the topic. In the course of the discussion it is not important to be the first speaker, but it definitely is important to make even and regular contributions throughout the discussion. This can be achieved when there is some one, who is willing to take up the rather tedious task of conjoining all the ideas and presenting them in a nutshell to the participants at regular interviews.

(iv) Listening: All this necessitates that the individual should possess capabilities of listening to what the other interactants are saying. Here, we must emphasize the difference between hearing and listening. Listening would only be evident when the listener shows signs of absorbing, assimilating and presenting the spoken material to the rest of the participants. On the contrary, if only hearing has taken place, it indicates that the speaker has not been able to penetrate the screen of indifference.

(v) Goal to fulfillment: These leadership qualities observed in the group, indicate that either the goal has been achieved or is in the process of being achieved. In other words, we can say that the topic has been thoroughly discussed by all the participants with appropriate input from the leader and all of them have been able to perform the important feat or task of bringing into focus the main/ancillary points related to the topic. Trying to conjoin the efforts of all the participants in a fruitful manner, would reveal all of them to be part of a cohesive group. This is definitely a Herculean task which only a leader could perform.

2. Knowledge of the subject matter:

Together with leadership qualities, the individual should also be well—read about the issues under discussion. His knowledge of the subject matter necessitates that, two things are kept in mind—the quantitative and the qualitative aspect of the topic. The quantitative aspect is concerned, figures and numbers should not be reeled off merely to prove one's point or knowledge about an issue, unless and until one is absolutely confident. There is bound to be someone who would be aware of the details. Keeping quiet, because of a lack of information or knowledge, is really not as bad as trying to impress the experts by spouting incorrect information. The quality of presentation would be an appropriate assessment of the topic and the issues discussed.

3. Analytical Ability:

The next is trying to present an acceptable picture of the self, is the capacity to use one's analytical ability to the optimum. While it is relatively simple to present data on an issue, it gives the appearance of a well thought out and analyzed presentation in the GD.

4. Clarity of thought:

Clarity of thought is extremely important and can be brought about by a distillation of the essentials and abandonment of the peripherals. The move in the discussion could be either from the core to the periphery or from the periphery to the heart of the issue. The participants should not get hooked to peripheral issues. There has to be concentration in the moves which would indicate awareness on the part of the participants regarding the subject matter and delineation of the topic.

5.   Conviction and flexibility:

Conviction: Whatever is being said should be stated with conviction. It often happens that the participants communicate their ideas in a group without really believing in them. This is more than evident at the face level and is easy for the experts to decipher and identify. It normally happens when the interactants harbours under the misconception that to be heard in the group is more important than positing of concrete ideas.

Often a participant makes an error in interpreting the topic which he realizes, much to his dismay, somewhere in the middle of the interaction. The need then arises to rectify the mistake and proceed along correct lines. The transition, which must be made, needs to be extremely subtle. Without really admitting that a mistake has been made, the speaker has an accept the view point of the other interactant and change sides to be one with those who have a more positive understanding of the topic.

(6) INSTANT PUBLIC SPEAKING

Meaning of Public Speaking:

An public speaking is a short talk on a set topic given to a tutorial or seminar group. In an public speaking one or more persons give a talk to a group and present views on a topic based on their readings or research.

Public speaking means delivering an address to a public audience. It also refers to public speaking and/or speechmaking. It is a brief discussion of a defined topic delivered to a public audience in order to impart knowledge or to stimulate discussion. Remember that the skill in public presentation is as important as effective writing. An effective public speaking should have an introduction, main body and conclusion like a short paper. Though it is a formal speech in nature or vocal performance to an audience, it may occasionally require adequate planning and thorough preparation in using one's voice, body language and visual aids such as slideshows to present and illustrate the points more effectively and to achieve the desired results.

Features of Public Speaking

A successful speaking

1.   Has a clear purpose;
2.   Addresses the intended audience at their level;
3.   Is well-organised, including a clear introduction, which motivates and previews the talk, and a summary;
4.   Avoids unnecessary details;
5.   Uses well-designed visual aids (and other) media;
6.   Engages the audience;
7.   Ends on time.

Format and Structure of Speaking
(A) Advance Preparation

Advance preparation and careful planning will prove helpful in making effective speeches. A checklist of the following steps should be kept in mind:

1. **Selection of a Topic:** The first and the most important task on the part of a speaker is the selection of the topic. The topic should be specific.

2. **State the Object:** The object may be one or the combination of two or more; such as to inform, to persuade, to influence and to entertain.

3. **Prepare an Outline:** The speaker should make a rough blue print like the introduction, body copy, main theme and conclusion.

4. **Narrow Down the Scope:** In a speech related to communication skills speaking skills, writing skills or listening skills should be specified. However, in some cases, narrowing the focus may be difficult.

5. **Locate Material and Data:** Collect information in books, documents, speeches, magazines, reports etc. and organise it in a proper from.

6. **Prepare a Rough Draft:** Prepare the rough draft and revise it. It should include an introduction, quotation, anecdotes, body, examples, conclusions and references.

7. **Use Audio-visual Aids:** Consider and procure Visual, Audio-visual Aids like charts, overhead projector, television sets etc.

8. **Rehearse:** It is necessary to recognise rehearsal and practice as important and advantageous for improvement in the delivery of speeches. Rehearsal provides self-confidence. It is desirable to rehearse in front of try-out groups. Inviting good points or bad points related to speech indicating verbal and non-verbal behaviour of the speaker is also important. This will help to inspire confidence.

A speaker steadily gains confidence, is able to anticipate questions from the listeners and answers the queries raised by the audience. The use of transitional phrases can be made to establish a relationship between ideas and concepts.

(B) Preparation

1. **Take Care of the Six Questions:** The speaker must ask himself the following questions. Once the speaker is clear in his mind about the answers to these questions, he will be able to have the desired effect. Given below are these questions that give rise to many other implied questions.

Question	Implied Questions
(a) What?	What do I wish to communicate? Have I thought about the content of my message? What facts and figures should I put forth?
(b) Why?	Why should the audience listen to me? Why have I chosen to speak to them?
(c) When?	Have I taken care of the timing of my speech? When are the listeners most likely to be interested? At what point of time should I say what they really would be interested in?
(d) How?	How can I best convey my message? Have I taken care to couch my message in the most persuasive language? Have I planned the beginning, middle and end of my speech? Do I need any audio-visual aids to make my speech effective?
(e) Where?	Where have I to speak? Or, in other words, what is the physical context of my communication? Will the audience be comfortable at the venue announced? Is the hall well-lighted and filled with adequate sound systems?
(f) Who?	Who am I going to speak to? Do I have to speak to an individual or several persons or to a large audience? What are the interests and expectations of the audience?

(C) During the Speaking:

1. **Be Clear and Organised:** Once the questions and their implications stated above have been taken care of, the speech will automatically turn out to be clear and effective. After all, clarity is the very life of all speech and writing. Moreover, it is also to be always kept in mind that no listener/reader likes to be caught-up in a jumble of confused thinking. It is basically a question of mental training and logical thinking that comes from all good education. Especially, in the world of business, clarity of thought is the greatest asset. All powerful speakers religiously stick to the principle 'Be Clear'.

2. **Be Simple:** From clarity of thought simplicity emerges. The simpler the language, the greater the appeal. A really effective speaker is one who can explain the most difficult or complex matter in the simplest language to a layman. Every educated person studies some subject in detail; likewise every worker having sufficiently long hands-on experience in a particular area becomes a kind of specialist. As a result, both the scholar and the worker pick up a special variety of language known as 'jargon'. No audience likes to listen to jargon. They can be patient only with the simplest language. Otherwise, they are likely to get bored and distracted.

3. **Furnish Concrete Details:** Many speakers spoil their speeches by talking in abstractions or making unoriginal remarks. An effective speaker, on the other hand, makes his speech vivid by furnishing details and actual experiences to capture the attention of the audience. Such gaining of attention towards the speech makes it full of life, it is the literal meaning of the word 'vivid'. It is therefore, very important for a speaker to make his speech lively and brilliant with eye-catching details, humorous anecdotes, relevant examples and enthusiastic eye-to-eye contact with the audience.

4. **Cultivate Effortless Grace and Naturalness:** Everybody knows that a cultured person moves and speaks with grace and sounds natural. But, then, there are circumstances in which it is really hard to act/ speak naturally. When one is face-to-face with a large, selected audience and has to make an important speech, it is quite natural to become conscious, perhaps over conscious and for the moment, to find it difficult to be natural. One has to devise his/her own ways to look, move and speak with grace. Practicing to speak in front of a mirror is a very common advice. Another is to find a popular television anchor/personality and emulate his/her ways.

5. **Enrich ones Mental Equipment:** Every effective speaker is a learned/well informed person. As the old Greek proverb says. "Out of nothing, nothing comes". So, one must remember that for every occasion and for every kind of speech, the best armament is a well-stocked mind – a mind stocked with all kinds of information, facts, figures, general awareness; readings in literature and philosophy; current affairs; economic and political developments; new advancements in science and technology; emergence of new business organisations and so on. Having the right kind of information for the right moment is an essential condition for speaking effectively.

6. **Be Brief:** Having a lot of information does not mean that one can go on rambling or going into unnecessary details, it means that one has to take care not only of the quantity of information but also of the quality of speech. It is said, 'time is money'. All that is superfluous must be cut out in order to make the speech concise. Every word one speaks is valuable and there are no words to waste.

7. **Be Informal:** The occasion may be formal but the speaker must strive to give his speech a personal touch. That is the only way to establish rapport with the audience and to create a lasting impression. Informality creates nearness.

8. **Be Enthusiastic:** Making an effective speech is not just a matter of doing a duty or performing a ritual. One has to get into the spirit of the occasion with enthusiasm and keen interest, only then will the speakers and the audience be able to empathise with each other. No one likes to listen to a dull or monotonous speaker. But an enthusiastic speaker gets an immediate response. Enthusiasm is contagious.

9. **Mind the Non-verbal Language:** Effective use of gestures is a necessary component of a speech, whether prepared or natural. In the same way, good eye contact is indispensable.

10. **Remember that Facts and Figures are not Enough:** If facts and figures had been enough there would have been no speeches. Anybody can have access to facts and figures. They can just be circulated or blandly stated. But a speech puts life into the dry bones of these facts and figures. They are just like a skeleton. It is the imaginative and effective use of language in a speech/writing that breathes life into that skeleton and supplies it with flesh and blood. Many great entrepreneurs, chairpersons, statesmen and scholars are known for their oratory that stems from their command of the language.

11. **Control your own Emotions while making an Emotional Appeal:** What comes from the heart goes to the heart. Every human being is full of emotions. But an effective speaker cannot afford to be carried away by his own emotions. On the other hand, while exercising poise and maintaining composure, a speaker can stir up his audience to action. His job is not just to inform but also to convince, influence and motivate his audience. That is how many speeches become memorable.

12. **Share Significant Experiences and Expertise with the Listeners:** It will not only give a personal touch to the speech to share significant earlier experiences and expertise learnt from them, but also give confidence to the speaker. It will also make the audience feel important to the speaker.

13. **Dressing Appropriately:** It is important to be dressed formally and appropriately to the occasion. A smart look could help boost ones confidence as well as the audience's confidence in the speaker. Wearing smart clothes also has an effect on the audience because it shows that the speaker respects them and one made the effort to dress appropriately.

14. **Introducing oneself:** One should begin a formal presentation by introducing oneself, even if most of the audience knows the speaker. It is important for everybody to know who the speaker is and which organisation he is representing as it makes the audience feel more comfortable when listening to the speaker.

15. **Using the Right Language:** Using formal but simple language, with some funny remarks every now and then, is one of the best ways to keep the audience interested and happy. Avoid too many details and many entertaining remarks, as this will make the audience feel distracted from the main topic.

    People are usually tempted to use complicated and very formal words when faced with a formal situation. Try to resist this temptation as it will confuse both the speaker and the listeners. This is particularly true in presentations where the contents can sometimes be hard to follow even when simple language is used.

16. **Talking with the Right Attitude:** The way one talks is a major contribution to the style of ones presentation. Although one should be oneself and talk as one normally does, one may tend to talk in an artificial way when put under pressure.

In order to reach the required degree of confidence and relaxation, it is necessary to train and prepare oneself before one faces the audience.

The style in which the matter is presented largely depends on the way one talks. Some presenters talk as if they are reading a newspaper, i.e., without any emphasis, pauses or body movements. This can grab the audience's attention for a short period, but as time goes by their interest in such presentations fades rapidly.

17. **Narrating Anecdotes and Humorous Events:** Presenting ones ideas in the same way as one would tell an anecdote to a friend is very effective in gaining the audience's undivided attention and trust. Telling a story involves a great deal of enthusiasm and a structured flow of events leading towards the conclusion. People enjoy listening to a story being told, whether it is about new company letterheads or the corporate strategy of an organisation.

18. **Speaking in a Moderate Speed:** If one talks too quickly, the audience may lose the link. On the other hand, if one is telling ones story at a slow space, the audience will be bored to death. Modernisation is the best opinion, as it keeps the audience interested and looking forward to hearing what is coming next.

19. **Using Pictures in the Slides:** Use slides to tell some of the jokes. A formal slide is not necessarily one which is full of words and complicated sentences. Funny sketches on the slides may be very effective in driving a point home as well as in entertaining the audience and winning their support.

20. **Thanking the Audience:** The speaker should also conclude by thanking everybody for being there and if applicable, answering any questions that they might have.

(D) After the Speaking

1. **Analyse and Introspect:** Generally a speaker tends to forget a bad presentation and remember good presentation. But a good speaker tries to analyse the reason behind the presentation being bad even if he is a senior speaker, hence amateurs have no reason but to analyse their presentation to find their mistakes and to overcome them.

2. **Methods to Overcome Flaws:** After analysing the flaws one has to find remedies to overcome them.

Common Problems with Public Speaking

1. **No proper Introduction:** Other people will probably not be as informed as you on the topic of your presentation, so be sure to provide a quick and simple introduction.

2. **Rehearsing is not done:** Proper rehearsing means half of the battle is won. To avoid common mistakes during presentation the following check is necessary:

   - Check that writing is legible from the back of the room.
   - Smooth transition between topics and slides.
   - Sequence of points is logical.
   - Get feedback from a classmate.

- Become familiar with the audio-visual aids.
- You shouldn't use a number of media (i.e. overhead projector and slides and blackboard) until you are quite confident and experienced. If you are going to use mixed media, it is even more important that you rehearse, to get an indication of how long it will take to turn off one projector, start the other etc.
- Timing of your presentation.
- Get used to public speaking and reduce your nervousness.
- Identify any mannerisms that may be inappropriate or annoying during public speaking. For example, these may include a tendency to begin every sentence with an ' errm ', or ' ahhhm ' or ' So. . ', or maybe you begin every new slide by saying "Where are we now?" or "Well, . . ". Once you have identified them, and with a little practice, you will be able to better control your use of these mannerisms. Don't worry too much about having a few of them in your seminars- you are not a robot! In addition, people generally have a very good ability to filter the occasional 'errm' or 'ahhhm' out of your main points. It is the overuse of such mannerisms that is problematic.
- **Remember:** It is far better to discover your mistakes and deficiencies in front of one or two classmates, rather than in front of a room that is full of students and staff.

3. **Being over-enthusiastic:** Keep your enthusiasm professional and objective.
Convince your audience that your subject is interesting, has important applications or is of great educational value.
If you are using overhead transparencies, use colour, clipart, pictures etc. to present (wherever relevant) maps, pictures of an organism, and pictures of a location.

4. **Conclusion is not proper:** Many speakers get to the end of a seminar, and no-one is sure whether they are finished or not. To avoid this, begin your last sentence by saying something like "Finally, I wish to say. . . ." or "In conclusion. . .", and then thank your audience for their attention. Do not ask the audience for questions. It is the chairperson's responsibility to thank you for your presentation. Then it is the chairperson who takes control of the question time and asks the audience for a question (NOT the speaker!).

5. No command over Language.

6. Too much of non-verbal communication.

7. Lack of subject knowledge.

8. Lack of knowledge for using audio-visual means.

Remedies to Overcome Problems of Public Speaking

1. **Stage Fear and Anxiety:** Everyone experiences stage fright, speech anxiety, or talking terror. The previous surveys by many researchers show that fear of speaking in front of groups is one of the greatest fears people have. Some techniques people use for coping with this fright is to try to motivate one-self that the audience understands our nervousness and know what we are feeling inside because they might have experienced this situation before hand and will honestly forgive mistakes.

2. **Nervousness:** The second fear and the most fears that people faced is nervousness. But be confident that the audience (some) will not notice the small changes in our voice or occasional mistakes but some will noticed even the small changes in our sound such as our pronunciation, the eye contact and the sound vibration. To avoid this thing, what we have to do is trying to practicing before the time. To make it looking perfect, doing the rehearsal many times is inevitable.

3. **Other Technique is being Ourselves:** Let the real us come through, relax and practice some deep breathing techniques. Other than that is beginning in our comfort zone. Practice with friends and share your fears with friends, check out the room first, the space, the equipment and the lights. Plus we should concentrate on the message, begin with a slow, well-prepared introduction and have a confident and clear conclusion. The most important thing is always be prepared and practice the script before present.

4. **Communication:** The problem of poor communication is complex and cannot be solved by a single book, a course, and certainly not by this short guide. The necessary steps include planning the presentation as with all academic undertakings. It is necessary to set aside quality time for researching the topic. Academic public presentations require depth and understanding of the material. It will not be considered of sufficient quality if unplanned presentation is given.

Make sure that the message delivered is successful achieved by the audience.

Exercise

1. What are soft skills? State the importance and elements of soft skills?

2. State the different types of speaking.

3. State the meaning of group discussion. What are the guidelines of group discussion.

4. State the process of group discussion.

5. What are the Do's and Don'ts of group discussion?

6. What are the different leadership skills?

7. State the features and format of public speaking.

8. Write short notes on:

    (a) Types of Etiquette

    (b) Importance of soft skills

    (c) Remedies to overcome public speaking

    (d) Overcoming stage fear

    (e) Public speaking

Part – VI

ETIQUETTES AND BODY LANGUAGE

(1) Telephone Etiquettes Listening / Speaking

(2) Problems of Telephonic Conversation

(3) Verbal/Oral Etiquettes

(4) Physical Appearance / Eye Contact / Body Language

(5) Group Discussion

(1) TELEPHONE ETIQUETTES LISTENING / SPEAKING

Phone Etiquette

Good manners are good for business, while great manners can make the organisation look great. This is true whether a manager is communicating face-to-face or on the telephone. Communicating over the phone is an integral component of conducting business. Today's sophisticated business requires that employees at all levels use the telephone for communication within the organisation as well as with outside stakeholders. This is more so as businesses are gradually moving out of brick and mortar facilities and people work from virtual offices, holding virtual meetings. More and more employees opt to work from home or from long distances even when they are on tours or vacations. They have to stay in touch with their peers and superiors. Secondly, businesses are getting dispersed, and outsourcing has becoming the order of the day. Suppliers and markets are spread across the globe. Decisions have to be taken fast. They can no more depend on the postal services, which is termed as 'snail mail'. Instant communication thus becomes crucial for the success of an organisation. Hence to be effective, it is essential that managers follow good phone etiquette and develop a pleasant phone culture. The simple guideline is: Treat your callers in a way that you would want them to treat you on a business telephone call.

Phone Strategy

- Treat the call as if it were a meeting - have a purpose, and an agenda.

- If the person you are trying to reach is not available, decide what you will do if someone else answers. Would you prefer to leave a message, go to voice mail, or hang up and call back later?

- If it is a scheduled call, be at your desk at the appointed time. If necessary, keep all the information and documents you may need while talking.

(6.1)

- Do not type or shuffle papers while you are on the phone - it suggests that you are preoccupied and are not listening to the caller.

- If you have to put the phone down, do it gently.

- If you have to call a number frequently, learn the names of the people who answer the phones.

Do's and Don'ts

- The telephone can catch the sounds that you make in your mouth. So, clear your mouth of food, chewing gum, cough drops, or any food before talking on the phone.

- If you have to sneeze or cough, turn your head and cover your mouth, and the receiver. Again, the receiver can amplify the sound and that may irritate the other person. Apologise and continue. If the cough continues, apologise, take a sip of water and continue.

- Cut down on the background noise when taking or making a call. Radios, televisions, and even computer noise can be distracting over the phone.

- Speak directly into the receiver. It should not be stuck in the neck or shoulder.

- If you dial a wrong number, explain yourself and verify the phone number so you do not repeat the call. Apologise and gently hang up.

Phone Etiquette

1. **First impression is good impression:**

   - Try to answer the phone on the second ring. Answering the phone too quickly can catch the caller off guard. Letting the phone ring longer would irritate the caller. It may indicate that the receiver has no concern for the caller's time.

   - Answer with a friendly greeting appropriate for the time of the day. Introduce yourself if the caller is known. If the caller is unknown, introduce your organisation and politely ask how you can help the caller.

   - Ask the caller for his name, even if it is not necessary. It gives the impression that the caller is important. It is a good practice to write down the name (ask for spelling if it is an unfamiliar name). This can be used in later conversations or while referring to the conversation in follow-up written communication.

   - While making a call, introduce yourself and politely ask for the person you intend to talk to.

   - Though the person is out of sight, always have a smile on your face, it gets reflected in the voice and tone.

   - Speak clearly and slowly. Never talk with anything in your mouth, including gum.

   - A telephone is a sensitive instrument. Hence lower your voice if you normally speak loudly. The entire neighborhood need not know that you are on the phone. It will disturb and distract others in the office.

   - Keep the phone two-finger widths away from your mouth.

2. **Call Transfer:**

   - If the caller asks for another person or department, do not transfer the call through the operator as the caller would then have to answer more questions or make further requests. It saves phone time and reduces irritation if the call is transferred directly to the desired person's extension.

- When transferring a caller, tell him who he is being transferred to. Inform that person that the call from the caller (announce the name) is being transferred to him. A brief statement of the caller's purpose may also be helpful in completing the communication

- **Putting a call on hold:** Ask other person's permission before putting him on hold. For example, if you need time to get a file and look for the relevant document, say, *"Would you mind holding while I get your file?"* While resuming the conversation, thank the person for holding.

- **Taking Phone Messages:** Include the following information while taking the call for someone else:

    - Caller's name and organisation name if applicable

    - Date and time of call

    - Purpose of the call

    - If the caller wants a return phone call

    - Phone number at which the caller can be reached

3. **Ending a call:**

- Do not close the call abruptly. Do not bang the instrument, as that is rude. Before hanging up, be sure that you have answered all the caller's questions. End in a pleasant by saying, *"Have a nice day"* or *"It was nice speaking with you"*.

- It is polite to let the caller hang up first. This shows that you are not in a hurry to get off the phone with him.

Handling business calls requires awareness of our unconscious action, which often amounts to bad manners, or lack of appropriate skills of effective invisible interpersonal communication. In telephonic conversation, the way we receive, respond, speak or hang up is often as important as what is communicated

Prepare Before Calling

Before initiating a call, be clear about the why (your purpose) and what (the exact content) of your call. For business calls, you must know who exactly you are speaking to and choose the most convenient time to speak to the person. You should also know whether you are calling as a follow-up action of some other earlier communication, or it is the first step in your transaction (of information and dealings).

- To be brief and concise, jot down the points you want to discuss. Always keep before you the points in the order in which you want to discuss them.

- Keep a (writing) pad and pen ready to note down any information worth recording.

- Consider whether the call is important from your point of view or from the receiver's point of view. In the latter case, you should structure your information from the receiver's point of interest. Begin first with what is important for your receiver. Talk about your interests later.

- Keep the duration of the conversation as short as possible. The other person may not be free to spare much time for your call.

How to Begin or Receive a Call

The first few words spoken by you as a caller or receiver are important for establishing your identity and purpose. They create the context for further conversation.

As a caller, you may not be personally known to the receiver. The receiver may be familiar with your purpose and your company, but may not exactly know you, unless you both have personally met and spoken to each other before. Therefore, begin with self introduction —your name, company, and purpose.

In the organisation, your call will be generally routed through a receptionist. The usual practice at the switch board is to attend to you within five rings. If you have been kept waiting longer than that, the receptionist should greet you with an apology. If your call is not answered even upto ten rings, it is advisable to cut off. Try some other number, if there is any.

Greet the receptionist with a "Good Morning." Tell your name and your organisation's name, "This is Pallavi Mehta of Nirali Prakashan" and then mention whom you want to speak to "Could I speak to Mr Ravi Chaturvedi, GM, HR, please?" Remember to be patient and pleasant while dealing with the secretary who is an important link between you (as a caller) and her boss (the person you would like to contact).

If the Call Is Cut Off

Many times, the telephone lines suddenly get disconnected. In such situations, courtesy demands that the person who originally initiated the call should redial immediately and say, "Sorry, the call got disconnected."

In case, the receiver has to suspend your call for attending to some other more important call, it is the duty of the receiver to resume the call and use some pleasant explanation to see that you do not feel slighted.

- Make your important calls polite by using words, such as 'please' and 'Thank you', when you make a request or get something done or completed.

- Always use the interrogative form for making a request. "Could I ...." Or 'May I ....". Use of direct categorical statement amounts to order.

- Even the statement, "I request you to connect me to so and so number/ or person" is not quite appropriate for requesting an unknown person to do something for you even if it is his/her duty. Instead, say "May I request you to ....".

Telephone Etiquette to be Observed:

In business, telephone calls are mostly received and passed on to the boss by the personal secretary. You have to be courteous with her/him. In return, she/he has to not only courteous but tactful too.

Sometimes, the secretary has to act very fast to find out whether the boss is free to talk to you or not available, before attending to you. Meanwhile, she/he knows what the boss wants. If the boss is present, but does not want to speak to you for some reason, the secretary will choose any one of the following polite things to say to you —

- "Sorry, he is busy in a meeting. May I have your number and he will call you later?"

- "He is busy with a foreign delegation. May I ask him to call you back as soon as he is free?"

These statements may not necessarily be true. However, they are intended to keep you satisfied even when the call is not successful. The secretary should never try to overhear the conversation between the two —the caller and the receiver (the boss). After putting through the call to the boss, she should go off the line.

Telephone Precaution

As a caller you do not know whether the person receiving your call is alone. Therefore, the first rule of telephone conversation is that confidential matters should never be discussed over the phone. They can be overheard/ tapped in transmission.

However, if you have to discuss something personal or something you won't like others to know, you should check with the person you are calling in a polite manner. For example, you may say, "Can we talk about the tender for the Golden Highway project?" or just ask, "Are you free? Can we talk about the tender?"

Communication over the phone requires the use of non verbal skills, such as pleasant tone, sweet voice, proper intonation, and clear articulation of words. We should be able to convey larger part of the message through our way of speaking rather than the meaning of words alone. Telephone etiquette involves good manners to create good business relations between two persons.

(2) PROBLEMS OF TELEPHONIC CONVERSATION

Phone Conversation: Most Commonly Used English Phrases on the Phone

Communicative skills are very important. Communicating properly on the phone is especially important, as the person you are speaking to cannot see your facial movement or your body language. They rely completely on what you are saying, and how you are speaking, to understand you fully.

As well as speaking clearly when talking on the phone, it is vital to use the right level of formality. If you are too formal, people might find it difficult to feel comfortable when talking to you. If you are too informal, they might think you are being rude!

Generally speaking, when you are calling in a business context (making calls related to employment, finances, law, health or applications of any sort), you should show politeness by using words like:

- *could*
- *would*
- *can*
- *may*

When making a request. When you ask for something, or receive help or information, you should use:

- *please*
- *thank you*
- *thank you very much.*

It is also okay to use some of the informal features of the English language such as short forms, phrasal verbs and words like okay and bye – in other words, everyday English! So phrases like:

- 'I'm off to a conference, okay, bye',
- 'Hang on a moment, I'll put you through'

are perfectly acceptable, as long as the overall tone of your voice is polite and friendly.

If it is more of an informal phone conversation (speaking to a friend, family member, close work colleague or even a friend of a friend), then a high level of formality is usually not required, but you should still speak with a polite manner, as it is seen as respectful.

It's fine to use less formal phrases in these conversations, such as:

* *'thanks'*
* *'cheers'*
* *'bye'*
* *'okay'*
* *'no problem'*

Another useful thing to remember is, it's better to ask for help or clarification when you're having a telephone conversation, than to pretend you understand something that you didn't. It is absolutely fine to use phrases like:

* *'Could you repeat that please?'*
* *'Could you speak a little more slowly please?'*
* *'Would you mind spelling that for me please?'*

Using phrases like these will help you to have a more successful phone call, and may save you from any problems later on. You could always say:

* *'I'm afraid the line is quite bad',*
* *if you can't hear very well.*

It also a good idea to practise words, phrases and vocabulary that you might need to use, before the call! So to help you out a little, here is a list of commonly used phrases:

Introduction / Making Contact

If answering a business call, start by introducing yourself or if the caller fails to identify themselves, then you could ask them to state who they are by using the following phrases:

* *Formal*
* *'Hello'*
* *'Good Morning'*
* *'Good Afternoon'*
* *'This is ___ speaking'*
* *'Could I speak to ___ please?'*
* *'I would like to speak to ___'*
* *'I'm trying to contact ___'*

Informal

* *'Hello'*
* *'Hi, it's ___ here'*
* *'I am trying to get in touch with ___'*
* *'Is ___ there please?'*

Giving more information

This would probably be used in a business context mainly, but could sometimes be helpful in an informal conversation too. It is good to specify where you are calling from, if you feel it may be helpful to the person you are calling.

Formal

- *'I am calling from* ___
- *I'm calling on behalf of* ___'

Informal

'I'm in the post office at the moment, and I just needed ___'

Taking / Receiving a Call

You may need to use these if you are answering someone else's phone, because they are unable to answer it themselves, or if you are answering an office phone.

Formal

- *'Hello, this is* ___ *speaking'*
- *'*___ *speaking, how may I help you?'*

Informal

- *'Hello, John's phone'*

Asking for more information / Making a request

If you need to ask for a specific person, then phrase your request as a polite question, if you only have an extension number and no name, you can say so. If you're calling for a specific reason, just explain briefly what it is.

Formal

- *'May I ask who's calling please?'*
- *'Can I ask whom I'm speaking to please?'*
- *'Where are you calling from?'*
- *'Is that definitely the right name/number?'*
- *'Could I speak to someone who* ___?'
- *'I would like to make a reservation please'*
- *'Could you put me through to extension number* ___ *please?'*

Informal

- *'Who's calling please?'*
- *'Who's speaking?'*
- *'Who is it?'*
- *Whom am I speaking to?*

Asking the caller to wait / Transferring a call

If you are transferring a caller to someone else, you should let them know that you are doing so, just so they know what is happening, as the silent tone could be mistaken for a disconnected line! If you are the one being transferred, you will often hear the person use the following phrases:

Formal

- *'Could you hold on a moment please'*
- *'Just a moment please'*
- *'Hold the line please'*
- *'I'll just put you through'*
- *'I'll just transfer you now'*

Informal

- *'Hold on a minute'*
- *'Just a minute'*
- *'Okay, wait a moment please'*

Giving Negative Information

If you are the one answering a call, you might not be able to help the caller. You can use some of the following phrases in these circumstances:

Formal

- *'I'm afraid the line is busy at the moment'*
- *'That line is engaged at the moment, could you call back later please?'*
- *'I'm afraid ___'s busy at the moment, can I take a message?'*
- *'I'm sorry, he's out of the office today'*
- *'You may have dialled the wrong number'*
- *'I'm afraid there's no one here by that name'*

Informal

- *'Sorry, ___'s not here'*
- *'___ is out at the moment'*

Telephone Problems

Telephone Conversations:

Having telephone conversations in a second language can be very stressful. If you don't know what to say, it is very common to feel nervous in any conversation. This is true even when speaking in your native tongue. One of the main reasons people get nervous is because they aren't prepared and know they might make mistakes during the conversation.

To improve confidence on the phone you must learn what to say. The first thing you should do to improve your telephone communication ability is to start out small by learning simple vocabulary and phrases. Start by knowing different greetings. It is so easy when learning English to try to do too much too soon and then get frustrated with not being able to speak as you had imagined. You have to start small, gradually developing skills and slowly working up to something more difficult.

Relax and enjoy yourself as well. Everyone knows learning a language can be frustrating! Don't worry if you make mistakes. Native speakers of English understand that you won't say everything the exact same way that they would. You shouldn't feel that you can't make any mistakes, no one expects you to be perfect.

In the following examples on English telephone conversations, we will give many examples of sentences and phrases you should know. From the start until the end of a telephone conversation we will go over everything all the way from greetings to goodbyes.

Here are a few sets of Telephone Conversations. Read the conversations in each set so that you will become familiar with the typical words frequently used in telephone conversations. Only over the period of time and after more practice hours, you could master how to converse in telephone. You might have had many such telephone conversations either in your social life or in your business life.

This is one of The Typical Telephone Conversations.

1. **Characters: Meena and Mohan**

   Mohan: Hello.

   Meena: Hello, Micro Computer centre.

   Mohan: Can I get Miss Deepa, please?

   Meena: Please hold on. Let me see whether she has come.

   Mohan: Hello, Deepa hasn't arrived. May I know who's calling?

   Meena: I'm Mohan, Deepa's brother.

   Mohan: Any message for Deepa?

   Meena: No thanks. I'll call again in the afternoon.

   Mohan: I'll tell Deepa that you've called and will call again. May I hang up?

   Meena: OK. Thanks.

2. **Characters: *Susee, Mani and Kumar***

   Susee: Hello! Is it Kumar's house?

   Mani: Yes. May I know who's on the other end?

   Susee: My name is Susee. I'm Kumar's friend. Can I speak to Kumar?

   Mani: Yes. Please hold on.

   Mani: Hello Kumar! What were you doing there?

   Kumar: I was watching T.V. Why did you call me?

   Mani: Your friend Susee is on the line.

   Kumar: Right. Let her be online for some time. I will be there in a minute.

   Mani: Hello Susee! Kumar is coming. Please be online for some time.

   Susee: OK. That is fine. I will be her waiting for Kumar.

   Kumar: Hai Susee! How are you? What is the matter?

   Susee: I am sorry to disturb you on Sunday. Could you come over here to meet our boss?

   Kumar: OK. I will be in our office in few minutes.

If you don't understand everything the other person is saying, be honest. Tell the other person immediately, otherwise you might miss some important information! Most people will appreciate your honesty, and will be happy to oblige.

Formal

- *'I'm afraid I can't hear you very well'*
- *'Would you mind speaking up a bit please?'*
- *I'm afraid my English isn't very good, could you speak slowly please?'*
- *'Could you repeat that please?'*

Informal

- *'Sorry, I didn't catch that'*
- *'Say that again please?'*
- *'I can't hear you very well'*
- *'Sorry, this line is quite bad'*

Leaving / Taking a Message

If the person you're calling is not available, be prepared to leave a message. This could be a voicemail, (which is a digital voice recording system), or an answering machine (this records messages onto a tape). If you're leaving a message with another person, they'll either ask if you want to leave a message, or you could request to leave a message with them. Be sure to leave your number, if you want the other person to call you back!

Formal

- *'Can I take your name and number please?'*
- *'Can I leave a message please?'*
- *'Could you please ask ___ to call me back?'*
- *'Could you spell that for me please?'*
- *'Can I just check the spelling of that please?'*

Informal

- *'I'll ask him to ring you when ___ gets back'*
- *'Could you tell ___ that I called please?'*
- *'I'll let ___ know that you rang'*

Saying Goodbye

The easiest part of the conversation! Simply be polite, and speak with a friendly manner.

Formal

- *'Thank you for calling'*
- *'Have a good day'*
- *'Goodbye'*

Informal

- *'Bye!'*
- *'Talk soon'*
- *'Speak to you again soon'*

Remember your manners!

It's very important to be polite on the telephone, use phrases like could you, would you like to, and to make requests, use please. Always remember to finish a conversation with thank you and good bye.

Write it down!

If you're nervous about speaking on the phone in English, then it may be helpful to write a brief script or a few bullet points on that you need to say.

If you will be speaking to someone you don't know, it helps to have things written down in front of you, to calm your nerves!

If you have a brief outline of what you need to say, it will help to organise your thoughts beforehand, and to use it as a reference during the call, if you get confused.

Phrasal verbs

One thing you could do to improve your telephone skills is to learn some of the phrasal verbs that are commonly used in English telephone conversations.

Common Phrasal Verbs

1. **hold on**

   means wait

   - *'Could you hold on a moment please?'*

2. **hang on**

    also means wait! (informal)

    *   *'Could you hang on a moment please?'*

3. **put (a call) through**

    means to connect one caller to another

    *   *'I'm just going to put you through now.'*

4. **get through**

    to be connected to someone on the phone

    *   *'I can't get through to his line at the moment, could you call back later please?'*

5. **hang up**

    means to put the receiver down

    *   *'I think the operator hung up on me, the line just went dead!'*

6. **call up**

    is to make a telephone call (mainly used in American English or slang)

    *   *'I'll call up the theatre, and find out about tickets.'*

7. **call back**

    is to return someone's call

    *   *'I'll ask him to call you back, when he gets home.'*

8. **pick up**

    means to answer a call / lift the receiver to take a call

    *   *'No one is picking up, maybe they're not at home.'*

9. **get off (the phone)**

    means to stop talking on the phone

    *   *'When he gets off the other phone, I'll pass on your message.'*

10. **get back to (someone)**

    means to return someone's call

    *   *'When do you think she'll be able to get back to me?'*

11. **cut off**

    to be disconnected abruptly during a telephone conversation

    *   *'I think we got cut off, I can't hear her anymore.'*

12. **switch off/turn off**

    is to deactivate (a cell phone/mobile phone)

    *   *'Sorry you couldn't get through to me. My phone was switched off, because the battery had died.'*

13. **speak up**

    means to talk louder

    *   *'I'm afraid I can't hear you very well, could you speak up a little please?'*

Hold on means 'wait' – and hang on means 'wait' too. Be careful not to confuse hang on with hang up! Hang up means 'finish the call by breaking the connection' – in other words: 'put the phone down.'

Another phrasal verb with the same meaning as hang up is ring off, but this isn't as commonly used as some of the other phrasal verbs listed above.

The opposite of hang up / ring off is ring up – if you ring somebody up, you make a phone call. And if you pick up the phone (or pick the phone up), you answer a call when the phone rings.

"Hang on a second..."

If you are talking to a receptionist, secretary or switchboard operator, they may ask you to hang on while they put you through – put through means to connect your call to another telephone. With this verb, the object (you, me, him, her etc.) goes in the middle of the verb: put you through.

But if you can't get through to (contact on the phone) the person you want to talk to, you might be able to leave a message asking them to call you back.

Call back means to return a phone call – and if you use an object (you, me, him, her, etc.), it goes in the middle of the verb: call you back.

Now you can start making those calls!

(3) VERBAL/ORAL ETIQUETTES

Every workplace evolves its own set of norms of behaviour and attitude. For example, you can survey banks or hospitals during lunch time. You may find that everyone resumes working without even a minute's delay, after lunch time, while in some others, taking an extra 10 to 15 minutes for lunch may be a general practice. In such cases, the sense of punctuality is governed by no rules written in the work manual.

Some business etiquette rules discussed here relate to the following:

- Giving introduction
- Telephone calls
- Business dining
- Interaction with foreign clients
- Inter-personal business etiquette
- Managing customer care

Here, an attempt is made to describe the behaviour and customs that would be considered appropriate and acceptable in most business organizations/offices/workplaces across the modern "educated" world. There may be some difference in the practices followed by different organizations.

The approach to business etiquette assumes that each business setting has its own business protocols, which an employee learns by working in that environment and by observing others. In an organization, the basic concern is to create a smooth work environment where each person helps others to carry out their jobs with ease. This is made possible by our ability to empathize with other workers' job concerns and priorities. This identification with others is the best form of business etiquette and culture.

Learning the rules of business etiquette will help us, as professionals, to act with ease in any business setting. Let us, therefore, consider some common situations in business and find out how to act appropriately.

Giving Introduction

Introducing Yourself:

A confident self-introduction always makes a positive first impression. But many people fail to do so. Either they think it to be a bold act, or they feel too shy to do it. But if you are going to meet a person for the first time, the other person is bound to feel comfortable to know who you are and why you are there)

Suppose two persons are waiting for an interview with the General Manager of a company. They are sitting in the waiting lounge across the corridor leading to the General Manager's office room. A well dressed middle aged looking person walks into the corridor and is standing near the Office of the General Manager. The candidates are not sure whether he is the concerned executive for whom they have been waiting. Now, suppose one of them stands up, moves up to him and says "Good Morning, I am Kashmira Seth. I am here for my interview with Mr. V. K. Grover." Hopefully the person would respond—"Good Morning! I am Grover. Pleased to meet you. We shall shortly have the interview." Ms. Kashmira Seth's bold self —introduction to Mr. Grover gives her an edge over the other candidate who remained silent. Most likely, Mr. Grover will have a positive and favourable impression of Kashmira, and construe that she is a confident, assertive, and enterprising young girl.

If there is an advantage in introducing yourself at the first opportunity, why do people shy away from doing so?

Many people treat introductions as a protocol to be observed as a ritual, when two or more persons meet formally.

Introducing yourself in a clear manner, pronouncing your first name and surname, when required, is necessary to help the other person know who you are.

Names, especially foreign ones, generally remain partially received, unless spoken out distinctly. Longer names have to be uttered slowly, in parts, so that the one person follows it fully.

Some Rules For Making Introductions Correctly

As a norm of business etiquette and the first step towards cordial business transactions, people greet each other by stating their full names and positions (in office) at the very outset.

1. Notice that first name and surnames are stated like "Dhiraj Rao, CEO, Sterling Gold informatics." not just "Dhiraj" or "Rao". Americans prefer to introduce themselves by their surnames only. Like "Lewis" or "Dickenson". But Britishers introduce by using first names and surname or "Tony Blair".

2. **Repeat your name when necessary:** During a conversation, the other party may forget your name or may have missed hearing it. At such moments you should help the host other party immediately by politely repeating your name "I am Irfan Mohammad, I am sorry, I should have told you.")

3. In the case of a prior fixed business meeting, if you are an expected visitor, first you should introduce yourself by telling your name and purpose: "I am Rajesh Bose and I have come here to meet Ms' Rakshanda Lahari in the Marketing Department. This is the normal practice followed by business executives.

4. Do not use honorific words, such as Sri, Smt, Mrs, Mr, Ms, or any other titles before your name, while introducing or referring to yourself. The other person can call you as Mr Kamat. If it is a degree earned by you, such as Ph.D., you may use 'Dr.' before your name and refer to yourself as "Dr. N K Gupta, or Dr. Gupta." But if you are a doctor by profession, people regard and know you as Dr. Naveen Gupta, or Dr. Hari Mehta or Dr. Roopesh Bhatia. But surgeons and physicians usually do not add the salutation before their names while referring to their names. This may be a universally appropriate self-announcement at the reception counter for registration. The point to be understood is that others may call us by adding titles or professional words or degrees, but we ourselves should not use them with our names.

5. Speak your name slowly and clearly. As mentioned, personal names sound unfamiliar. Therefore, they should be articulated as distinctly as possible. If required, help others by spelling your name.

Handshake

As a winning form of non-verbal communication, handshakes must be accompanied by eye contact and a gentle smile.

Today most business meetings begin and end with a handshake. A handshake is immediately done after introduction by extending your right hand and firmly holding the other person's right hand very briefly. In modern business a handshake is a non —verbal clue of friendly dealings.

As a visitor, you can first offer your hand for a handshake with your host. In fact, the handshake is so spontaneous that both the parties almost simultaneously put forward their right hands towards each other. Sometimes, while parting, people shake hands by holding both hands together and putting and their left arm on the back or the shoulder of the other person to communicate the warmth that has developed between them after meeting each other.

In some situations, you may express your feelings by saying —"Pleased to meet you". And the other generally responds —"My pleasure" or "So, am I". But these words are just pleasantries. They do not mean much as verbal communication.

On Failing to Recall Someone's Name

In business, we may have been visiting a person often. It may be possible that on meeting the same person somewhere else, we do recognize him/her by face, but fail to recall their name. Before the other person detects your failure, you should ask for his/her card. You can say "Could I have your latest business card for your telephone number/and address?" or you can just say, "Could I know your full name? I have your initials."

To be tactful in such situations is also good verbal etiquette. If you let the other person know that you have forgotten his/her name, it may make the person feel that he/she is not important enough to be remembered (by name). So, we should act as if we know the name, but we wish to have more details about the person.

How to Introduce other Persons

One of the difficult things you have to do is perhaps to introduce other persons at a business meeting. We should know each other in terms of their names and professional status before we conduct the meeting or discuss the transaction. The status communicates the role the person plays in the business transaction.

The Protocol (Rule) for Introducing Others

1. Normally, the senior most people among the visitors, or the host team, introduces the other members of his/her group present there.

2. The practice is that first visitors are introduced to the host. Then, members of the host group are introduced.

3. The rule is that we do not introduce a senior to a junior. Instead, we always introduce the lowest ranked person to the highest ranked person. Always say "Mr. Sharma. (Chairman CMC), may I introduce Payal Swaroop to you? Payal is this year's university topper and Gold Medalist, working in our placement department."

    Notice two things here:

    One, the polite form "May I introduce ..." is appropriate and formal, when you are speaking to a superior. But to others you can just say, "This is Shikha Gulati. Shikha is a senior lecturer in finance."

    Two, we repeat the name so that it is duly received and remembered by the other person. But to repeat the name naturally, we create a context by mentioning the most significant detail about the person concerned.

4. Suppose there are just two persons to be introduced to each other. As a rule, you should first introduce the junior to the senior, as stated above. After introducing the junior to the senior, you introduce the senior person to the junior person.

(4) PHYSICAL APPEARANCE / EYE CONTACT / BODY LANGUAGE

(a) Proxemics (Perception of Physical Space):

There are several non-verbal signals which are understood by the sender and the receiver. One such source is space. It communicates in its own way. *The degree of perceived physical or psychological closeness between people is called proxemics.* The term proxemics was introduced by anthropologist Edward T. Hall in 1959 to describe set measurable distances between people as they interact. Communication experts call this dimension of communication, proxemics. It can also be called as 'space language'. In this context, space means distance between the sender of the message and the receiver of the message in oral communication. Space can be classified into various categories according to distance.

1. Intimate (soft whispers, pat on back etc.)
2. Personal (close circle, shake hands etc.)
3. Social (used in official relationships)
4. Public (formal and objective)
5. Fixed and Semi-fixed (arrangement of furniture).

- Broadly, Intimate (0-2 ft.), Personal (2-4 ft), Social (4-12 ft.) and Public (more than 12 ft.). When strangers enter the wrong zone, we feel uncomfortable. E.g. In an airplane / elevator, this space gets compromised and we deal with it by "dehumanizing" (making no eye contact and acting busy) those around us. Even a loud cell phone conversation trespasses this space!

Business Relationships:

- Generally, business relationships start in the Social Zone, but as the relationship develops and trust is formed, they move to the Personal Zone.

Business Seating:

- In competitive situations, people sit across-the-table, monitoring each other. When working independently, they sit across but never directly opposite, as this provides adequate isolation and privacy. In a group, the leader always sits at head of the table, as this position brings with it a sense of power.

Cross-Cultural Variance:

- Personal zones vary across cultures, professions, personal preferences and affluence (more affluent persons demand more personal space). Also, those living in densely populated environments (India) tend to have smaller space requirements. Generally, low-context cultures (North American, Northern Europeans) prefer interacting in the Social Zone for business but high context cultures (Mediterranean, Arab, Latin) favour the Personal Zone.

- People value their personal space and business leaders use this space to "read" the non-verbal messages. The trick lies in being sensitive to and respecting the space of others, leading to working relationships filled with trust!

- The physical environment around us has its own non-verbal language. Surroundings can include colour, design, layout and time. They play an important role in determining non-verbal signals. On a more universal level, smiling is a good example used by all cultures, which is viewed in the work place as a way of signaling friendliness and co-operation.

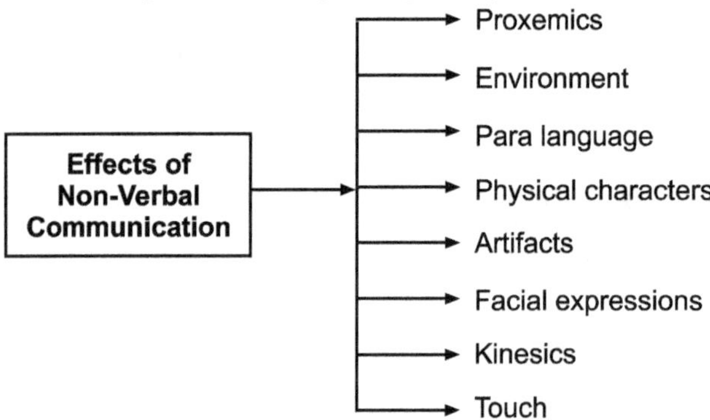

Fig. 6.1: Effects of Non-verbal Communication

In brief, it can be stated that the dynamics concerned with the distance between the speaker and the listener is termed as proxemics. Distance affects interpersonal space. Personal space is an invisible factor.

(b) Chronemics (Time Language):

- One of the most immediate features of chronemics is the attitude of different cultures to punctuality. Different cultures give different weightage to punctuality. For example, consider a multinational organisation where, managers from different countries are required to attend a meeting at a specified time. There will be cross cultural communication in such a gathering. Since each country has a different idea and/or concept of punctuality, all the participants would arrive at different times, some would arrive before time and some even after the specified time. None of them would pay much attention to either arriving early or late probably because of the different cultural attitudes to the time factor or punctuality. As in the case of punctuality, there is also a varied cultural attitude towards management of time and orientation to time with respect to past, present or future. Therefore, there is a problem about how cultures formulate their approach to time. A skilful management of activity, therefore, depends on one's attitude towards synchronic and sequential time.

- Under time language, indications are given about how important time is. Nowadays, time management has become an important part of business transactions. Organisations adopt various means and methods to save time because after all 'Time is money'. All communication should be appropriately adjusted to be effective.

- Use of time is a very subtle factor in non-verbal communication. Time is non-verbal communication. Having someone give you time to speak, or giving you their time and listening to what you are saying, can greatly boost self-esteem. A person who wastes his own time and that of others can certainly be considered inefficient. Chronemics includes a sense of time used for presentations, conducting meetings, giving good or bad news etc. as it creates good and bad impressions. In brief, it can be stated that chronemics is an important non-verbal method of communication as time also conveys certain messages.

(c) Artifacts:

- Artifacts are concerned with the environment and objects around us. The physical environment around us conveys its own non-verbal language. Surroundings consist of different objects. There are two important parts of surroundings - colour, and layout. Colour suggests different attitudes, behavioural patterns and cultural backgrounds. For example, white colour stands for peace and surrender and red suggests danger. Decorative articles are also useful in effective communication. Artifacts are also personal objects we use to announce our identities and personalize our environments. Artifacts are thus material objects as an extension of oneself. For example, clothing has the power to influence. Change left in a phone booth was returned to well dressed people 77% of the time, poorly dressed people only 38% of the time.

- Several studies show that fancy suits, uniforms and high-status clothing are related to higher rates of compliance.

- Artifacts announce professional identity. Nurses and doctors wear white and often drape stethoscopes around their necks; executives carry briefcases, whereas students more often backpacks. White-collar professionals tend to wear tailored outfits and dress shoes, whereas blue-collar workers often dress in jeans or uniforms and boots. The military requires uniforms that define individuals as members of the group. In addition, stripes, medals, and insignia signify rank and accomplishments.

- We also use artifacts to define settings and personal territories. When the president speaks, the setting usually is decked with symbols of national identity and pride such as the flag. At annual meetings of companies, the chair usually speaks from a podium that bears the company logo. In much the same manner, we claim our private spaces by filling them with objects that matter to us and that reflect our experiences and values. Lovers of art adorn their homes and offices with paintings and sculptures that reflect their interests. Professionals may decorate their offices with expensive furniture and framed awards to announce their status or with pictures of family to remind them of people they cherish. Web sites, too, are defined by artifacts such as photos, animations, and colours that express the creators' personalities and interests

- Although the human body transmits a number of messages, these messages do not carry any sense until they are translated by someone. The process of translation is important where artifacts are concerned.

- In order to communicate by means of embodiment media, both the sender and receiver of the message must assign the same meaning to the common embodiment media. The assignment emerges through interaction between humans and artifacts. The designer of an artifact extracts essential characters of human communication by observation and it is to be implemented in the artifact. All artifacts do not have the same character (embodiment) as humans.

- "No" is often represented with a shake of the head. In order to express a negative message, the sender must have appropriate embodiment media. It means that the sender must have a head, neck, body etc. and the ability to realise the action. Lifeless articles like a chair, table, briefcase do not have such embodiment. Human beings can adopt various other methods to understand embodiment media.

Aspects and Types of Body Language (Kinesics)

- There are various ways in which non-verbal communication can be classified. There are four most common types:

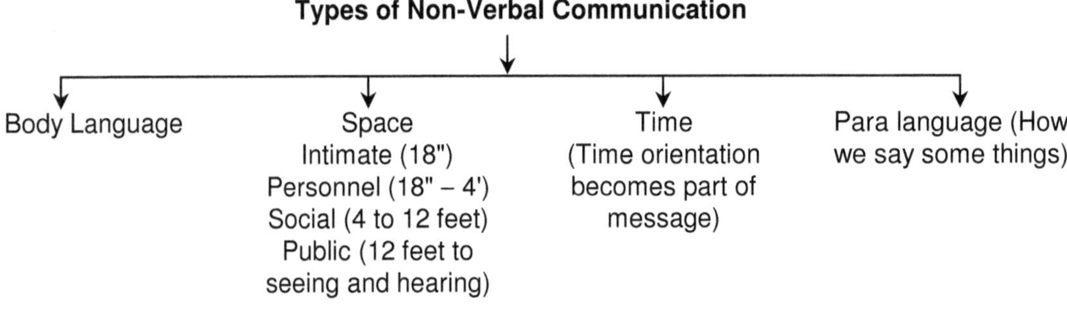

Fig. 6.2

- These four types are important in speaking and listening. The human body sends non-verbal messages through arms, fingers, expressions, postures and so on. Face and eyes are most important. Gestures are useful for sending non-word messages through body parts. Another area of body language is physical appearance; for example clothing, hair style, jewelry, cosmetics etc. can affect body movements. Yet another type of non-verbal communication includes space and the way it communicates, meaning in speaking and listening.

Kinesics:

- The study of body motion and facial expressions as related to speech is called *kinesics*. It is an important factor in non-verbal or oral communication.

- Kinesics is non-verbal behaviour related to the movements of a part of the body or the body as a whole. Generally, all communicable body movements are classified as kinesics. Kinesic communication is one of the most important non-verbal communication forms. Communication by way of body movements is endless. For example, when the forefinger and middle finger are indicated in the shape of the letter 'V' it becomes the sign for victory. In certain cases two fingers are used to symbolise number two. Sometimes '0' may be indicated as OK, in other cases it is taken as a zero.

- There are some non-verbal signs that regulate and maintain the flow of speech during meetings or conversations. These can be both kinesic, such as nodding of the head, as well as non-kinesic, such as eye-movements. The information which flows as a part of feedback to indicate whether the receiver has understood the message can be highly confusing.

- Kinesics is an important part of non-verbal communication. The movement of the body conveys a specific meaning and interpretation. There are movements which carry a significant risk of being misinterpreted in certain cases of communication.

- Body language is a comprehensive term used for the method of communication through different parts of the body other than the tongue. It includes, all sorts of body movements, gestures, facial expressions, postures, miming, touch etc. depending upon the physical space between the communicators. Silence as a body communication also plays an important role in certain exchanges. Body language, by and large, is unintentional and more truthful than verbal messages. It is easy to deceive a person by expressions.

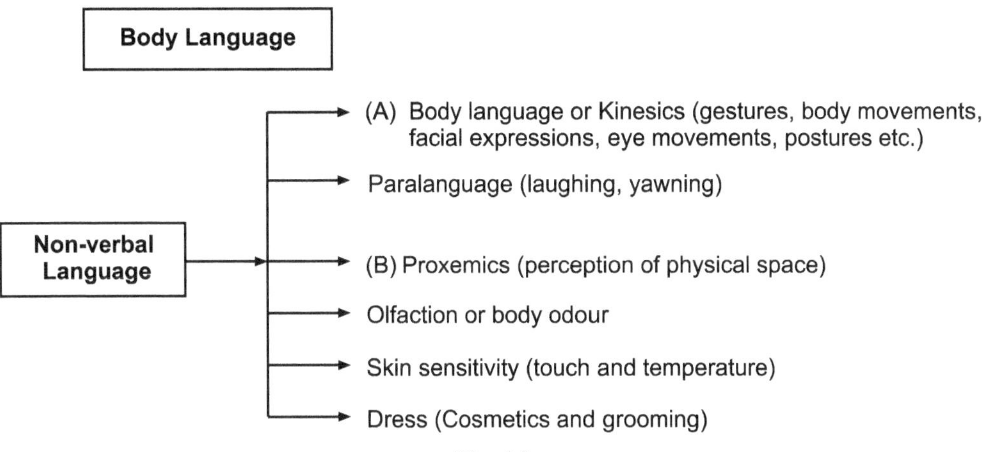

Fig. 6.3

Presentation Plan

- The art of public-speaking is the fruit of constant efforts. There is no definite formula to be adopted to become an effective speaker. Enough planning and preparation is essential for a successful presentation. Preparation is the key to overcome nervousness.
    1. As far as possible the speaker should not read out a written speech word by word.
    2. Written speech seldom sounds fresh.
    3. It is not necessary to memorise the speech either. It never exhibits spontaneity.
    4. Memorisation of speech would hamper flexibility. Consequently, communication will suffer.
    5. Face to face conversation or interaction involves thinking and speaking and not mere repetition.
    6. Even while reading the speech, one must lift the head occasionally and observe the audience.
    7. The speaker must make adequate notes and points and arrange them in a proper order before actually starting the speech. It should bear proper headings and sub-headings.
    8. Appropriate time should be devoted for introduction and to the main body of the talk.
    9. Notes and points should be written neatly and in bold letters.
    10. A positive attitude needs to be generated while speaking.

(a) Facial Expressions:

- Facial expressions exhibit attitudes. A smile expresses friendliness and affection. Raising of eyebrows conveys surprise, while a furrowed forehead expresses worry and anxiety. Various parts of the human body, particularly eyes, are used for communicating information. Eyes are the most significant organs which communicate subtle messages. Really speaking, eyes by themselves do not show emotion; but when they are associated with other parts of the face, they are windows to the soul. Squinting, winking, staring and gazing express a world of meaning. Looking at the eyes, face and other verbal signals during conversation perform certain important functions:
    1. Looking helps with the synchronising of utterances.
    2. It provides definite feedback.
    3. It provides additional non-verbal information which elaborates verbal messages.

(b) Eye Contact:

- Eye contact is an instantaneous and effective means of establishing rapport with an audience. A good speaker first looks at the audience and takes a pause before beginning his speech. It helps him to make a good impression on the audiences. In case the speaker gives breaks and puts his ideas into small units rather than in a lengthy narration, it allows listeners some intervening time to grasp those ideas. Once this is done, the speaker can proceed to explain the subsequent points or ideas. While speaking it is also necessary to maintain eye contact with the audience. If the listeners feel that the speaker is ignoring them, they are also likely to ignore the speaker and what he says. Through eye contact, the speaker gets signals whether the channel of communication is open and operating on right lines. This sense itself works as the feedback.

(c) Vocalics (Aspects of Speech or Para Language):

- The dictionary meaning of vocalics is 'concerned with a vowel/vowels'. Vocalics is the study of non-verbal cues of the voice. Vocalics means such features as are related to voice, including intonation and pitch. It is the area of language related to the way voice and intonation are used. Voice has characteristics like tone, volume and pitch. Tone can be considered as the quality of the voice. Volume is the loudness or softness of speech which can be suitably modified according to space, distance, number of listeners, nature of audience and self-confidence etc. A high pitch voice is not preferred because it gives an impression of immaturity. Voice generally becomes high pitched when a person is encompassed by a sense of fear. Rapid speech shows some kind of excitement. Stress on a particular syllable can alter the meaning of a sentence.

(d) Gestures:

- In certain cases, gestures are used extensively when a person does not know the local language. The deaf mainly depend upon gestures for their sign language. In certain discussions, gestures of a person convey much more than what he speaks, for e.g. enthusiasm, attitudes, emotions, etc. Gestures like shaking the hand display friendship and warmth. Other such examples are holding of hands, patting on a shoulder etc.

(e) Postures:

- Each movement or posture of the body has expressive and defensive functions. The way in which one sits or stands, walks in or walks out tells a lot about the person's emotions and attitudes. But no body movement itself has any exact meaning. Just as spoken words are assisted by gestures, posture is also taken together with other verbal and non-verbal clues. For example, lowering of the head indicates the end of the statement, or raising of the head indicates the end of the question. A large number of postural changes indicate the end point in interactions; for example, the end of a thought or the end of a statement. A shift in posture shows that something is happening. This aspect of body language is known as "Proxemics."

(f) Dress and Appearance:

- The kind of dress that is worn by a person and his grooming shows his status and attitude. Most people are influenced by how the speaker looks and the clothes he/she is wearing. Physical attractiveness plays an important role in one's assessment of people. Attractive people are generally considered to be more successful, more intelligent and more competent than those who do not look or appear attractive. A person's physical appearance and dress conveys a great deal of information about him. One's judgement or evaluation of others is greatly influenced by physical appearances. It has a definite impact on the communication process.

(g) Haptics (Communication by Touch):

- Haptics is a scientific term that indicates the sense of touch. Haptics is the science concerning human interaction by means of touch. The word haptics is derived from Greek, meaning *"to grasp, touch or perceive"*. Haptics is often accomplished through electrical motors. There are other methods to create pressure sensations such as those with devices that are hydraulic or pneumatic etc. In feedback, a user can feel forces applied directly to the skin, which are detected by the user through sensors within the skin. Tactile feedback can also be applied by the user through electrical currents applied directly to the skin or objects that can vary in temperature touching the skin. Haptic devices have varying complexities and can work in different ways. Pressure devices are often described by their degrees of freedom. Haptic devices are available for commercial applications.

(h) Voice Modulation:

- The speaker must know how he should use his voice effectively. To want to become an effective speaker, a good voice is a must. A good voice is no doubt a natural gift; but one can make efforts to improve the quality of one's voice with skilful training and practice. Proper attention to this aspect would lead to the acquisition of confidence and power.

    1. Variations in the pitch and tempo of the voice are essential to transmit the message effectively.
    2. Fast delivery of speech indicates lack of confidence, and betrays the object of the speech and the image of the speaker. The speed of delivery must be such that the audience can keep pace and understand the matter properly. One should not go beyond the speed of 125 to 150 words per minute. One has to bring the speed of speaking within this range by proper practice.
    3. Pronunciation of the words used must be correct with due stresses wherever required.
    4. One should speak loudly enough so that everyone including even the last man in the audience is able to listen to the speech clearly.
    5. Do not use repeatedly words like "You see", "I mean," "Do you understand," "Is it clear" etc.
    6. The microphone should be kept eight to ten inches away from the mouth to ensure proper transmission of ones voice to the audience.

(i) Audience Awareness:

    Before starting the speech the speaker should first:

    1. Size up the listeners.
    2. Consider the age, sex, background and interest of the listeners.
    3. Ascertain whether the audience is friendly or hostile.
    4. Use common sense to become a good speaker.
    5. Select an approach that suits the audience.
    6. Make the listener feel that he/she is being addressed individually.
    7. No verbal fireworks are necessary to arrest the attention of the audience.
    8. Make the audience feel that the speaker is sincere in gaining their interest.
    9. Dramatise certain ideas to overcome the barriers of communication.
    10. Create an impression that you want to share the views and ideas of the audience.
    11. A 'joke' should be delivered in such a way that it appears to be part of the speech. The audience should receive it almost unprepared.
    12. Do not get distracted if a listener smiles or whispers to a neighbour.
    13. Concentrate on ideas.

(j) Silence:

- *Speech is silver, but silence is golden.* It means silence speaks louder than words. It is a non–verbal language par excellence. It establishes the relationship between communicators and their attitudes towards one another. Moments of silence show that the parties involved are either confused or do not know how to continue the conversation. In case of public speaking, when the speaker punctuates his remarks with periods of silence, he could be using it for stressing the importance of his message. In some cases, silence is generalised. For example, when a worker fails to complete his assigned job and the supervisor makes enquiries about it, his silence speaks for his failure.

(k) Physical Appearance:

- How one looks and what one wears affects listeners. Thus some organisations are more causal in dress, others are not so. Learning an organisation's culture includes learning its dress code. Most organisations encourage men to wear a shirt, a tie and a suit and women to wear a white blouse and a dark skirt and a jacket. But these rules are continuously changing. Terms as 'smart casual' or 'business casual' indicate symbols for appearance, and it is important that one knows what such terms mean is one's particular surroundings or organisation.

- Clothing, hairstyles, neatness, postures, stature are part of personal appearance. They convey impressions regarding occupation, age, nationality, social and economical level, job status and good or poor judgement, depending on some conditions. Aspects of surroundings include room size, location, furnishings, machines, architecture, wall decorations, floor, lighting, windows, view and other related features wherever people communicate orally. Surroundings will vary according to the status and according to the country and culture.

- Whether one is speaking to one person face to face or to a group in a meeting, physical appearance and surroundings convey non-verbal stimuli that affect attitudes towards spoken words.

Professional Dressing

Objective language is non-verbal message communicated through appearance. As a medium of non-verbal communication, it includes arrangement and display of material things. This method may include intentional or unintentional communication through material things like clothing, make-up, ornaments and other accessories, books, buildings, room furniture, interior decorations etc. These offer signals relating to the context, such as formal and informal settings.

Dress and decoration communicate a great deal about the speaker's rank, status, personality, feelings, emotions, attitudes, opinions etc. What they reveal is something special about the person. For instance, when a professional bathes every day, wears clean clothes and his socks are not stinking, it indicates his outlook and concern for others. On the other hand, if he has body odour and wears clothes and socks which have not been washed for a couple of days, he will be looked down upon right at the outset. First impressions count and whether one likes it or not, people are immediately judged by their appearance.

Professional dressing conveys a non-verbal message as it creates an impression about the person, his background, his tastes and choices. Obviously, the person is instantaneously assessed by the audience. Hence the emphasis is on choice of dress and accessories.

Professional dressing includes dressing appropriately for a job, an interview, an internship, a networking event, etc. Each company and event is different. Secondly, there are many other variables, like time of event, season, the city of residence and work, mode of transport being used, and whether person will be sitting or standing most of the time, dress code announced, what the others are likely to be wearing etc.

Types of Business Attire: There are three types of dress codes viz., Professional Dress, Business Casual and other Casual.

1.   Professional Dress: Professional dress is the most conservative type of business wear for most 'white collar', job situations, whether it is accounting, finance, or other conservative industries. For men, professional dress means a business suit or a blazer, dress pants and a tie. For women in the Western societies, this means a business suit, pants suit, or dress and jacket. However, in India, one can see even ladies holding top level jobs wearing saris. At the worker level, viz., 'blue collar' jobs, there are the uniforms, overalls and safety gear.

Fig. 6.4

For those in specialised services like the defence services, police forces etc., there are the Service-Regiment-Seniority-specific uniforms, which instantaneously communicate the status of the person. Religious leaders, political leaders, lawyers, judges, doctors, nurses, sports persons, porters, security persons and other workers have their own way of dressing, which identifies them distinctly.

The professional dress worn by people of different countries varies. What is formal office wear by women in India may be considered informal and dressy in the western countries. Many companies have an established dress code which is binding on the employees.

A professional who does not take the time to maintain a professional appearance presents the image of not being able to perform adequately on the job. If a person looks and behaves like a highly trained and well-groomed professional, he/she will win appreciation, respect and honour. Apart from dress, professional men should take care of various aspects of grooming, such as a fresh haircut, removal of all facial hair, nicely shined shoes and a crisp, nicely pressed suit etc., as they go a long way in establishing a professional demeanour. Cologne and after-shave are optional, but if used, it must not be very strong.

Women professionals as a rule do not wear heavy jewellery, very strong cosmetics and perfumes etc., and they keep their hair neatly organised. Tattoos are discouraged in many companies in America. Those who display them are considered unprofessional, low-class and ignorant.

There are a number web sites where the 'do's & 'don'ts are described.

2.   Business Casual: This is a more relaxed but not casual version of the 'Professional Dress'. If the organisation environment is semi-conservative, wearing business casual would be appropriate. Apart from this, some interviews and events may also call for business casual.

Basically, for women, business casual is a shirt with a collar and/or a sweater, khakis or dress pants and nice shoes. They can also sometimes wear a moderate length (knee length or longer) dress or skirt. For men, business casual is a polo shirt or shirt with a collar and/or sweater, khakis or dress pants and dress shoes. No tie is required.

3.   Other Casual: This is a type of comfortable outfit a person wears on an everyday basis, or for casual networking events in the organisation, but does not to work. However, It includes jeans, tee shirts, flip flops, sneakers.

4.   Adornment and Artifacts: The way in which people carry cigarettes, pipes, canes or glasses also suggests different semiotic meanings. (Semiotics is the science of the emotional or psychological impact of signs, appearances and of how things look). This understanding is gaining a lot of importance, especially in audio-visual communication including advertisements and entertainment, where a lot has to be communicated very effectively in a very short time.

(5) GROUP DISCUSSION

Group Discussion : Refer to Point (3) of Chapter No. 5.

Group Discussion is not a debate in which you either oppose or support the topic. There are no clear cut positions or stands to be taken. GD is a continuous discussion, a live interaction, in which you examine a subject/problem from different angles and view points. As a participant, you may disagree with or support the other's point of view or bring in a new point of view. This should not be done by showing disrespect for the other person, even if you do not accept his/her point of view. Courtesy in discussions indicates our level of culture and sophistication.

Some techniques or guidelines to be followed by the Group Discussion participants are as follows:

1. **How to join the discussion:**
   - I'd like to raise the subject of ......
   - What I think is ......
   - In my opinion ......
   - If I had to say a word about it ......
   - I feel strongly that ......
   - May I make a point about ......

2. **To support what some other participant has said:**
   - I'd like to support the view point of Mr. A about ......
   - I completely agree with Mr. B about the point ......

3. **To support disagreement:**
   - I would like to offer a different viewpoint ......
   - Please allow me to differ here ......
   - I think differently on this issue ......
   - I do not agree here, in my opinion ......

4. **To make a point very strongly:**
   - I am convinced that ......
   - You can't deny that ......
   - Anybody can see that ......
   - It is quite obvious that ......

5. **To bring a discussion back to the point:**
   - Perhaps, we could go back to the point ......
   - Could we stick to the subject, please ......
   - I am afraid; we are drifting away from the point.

Your analytical ability and your verbal and non verbal skills of communication give you a competitive edge over others.

Listening in Group Discussions:

In Group Discussions, only speaking is not essential, but listening has its own vital importance. Only a good listener can be an effective speaker and can bind and convince the group with his opinions. In GDs, Listening is also a participative act. Listen to what others have to say. Do not listen with a desire to refute. Listen to assimilate and analyse; then speak to express your thoughts in the light of thoughts of others.

Do not interrupt, but try to join in the discussion tactfully.

Finally, if you really want to stand out, do not try to dominate by demolishing other participants.

Exercise

1. What are the different etiquettes of telephone?
2. What are the Do's and Don'ts of telephone?
3. What are the different aspects and types of body language?
4. Write short notes on:

   (a) Physical appearance   (b) Telephone etiquettes
   (c) Facial expressions   (d) Eye contact
   (e) Body language   (f) Artifacts
   (g) Postures and Gestures   (h) Silence
   (i) Audience Awareness

MODEL QUESTION PAPER

1ˢᵗ Semester Diploma Engineering Examination, 2018 (New Syllabus)
Communication Skills

Time: 3 Hours   Full Marks: 80, Pass Marks: 26

1.   Read the following passage and answer the questions given below: [2 × 5 = 10]

An e-reader is a device that allows you to read a book-length publication in digital form, consisting of text, images, or both, and produced on, published through, and readable on computers or other electronic devices. Sometimes the equivalent of a conventional printed book, e-books can also be born digital. The Oxford Dictionary of English defines the e-book as "an electronic version of a printed book," but e-books can and do exist without any printed equivalent.

E-readers are superior to printed books because they save space, are environmentally friendly, and provide helpful reading tips and tools that printed books do not. E-readers are superior to printed books because they save space. The average e-reader can store thousands of digital books, providing a veritable library at your fingertips. What is more, being the size and weight of a thin hardback, the e-reader itself is relatively petite. It is easy to hold and can fit in a pocketbook or briefcase easily.

In addition, e-readers are superior to books because they are environmentally friendly. The average novel is about 300 pages long,. So, if a novel is printed 1000 times, it will use 300000 pieces of paper. That's a lot of paper! If there are about 80000 pieces of paper in a tree, this means it takes almost 4 trees to make these 1000 books. Now, we know that the average bestseller sells about 20000 copies per week. That means that it takes over 300 trees each month to sustain this rate. And for the super bestsellers, these figures increase dramatically. For example, the Harry Potter book series has sold over 450 million copies. That's about 2 million trees! Upon viewing these figures, it is not hard to grasp the severe impact of printed books on the environment. Since e-readers use no trees, they represent a significant amount of preservation in terms of the environment and its resources]

Finally, e-readers are superior to books because they provide helpful reading tips and tools that printed books do not. The typical e-reader allows its user to customize letter size, font, and line spacing. It also allows highlighting and electronic bookmarking. Furthermore, it grants users the ability to get an overview of a book and then jump to a specific location based on that overview. While these are all nice features, perhaps the most helpful of all is the ability to get dictionary definitions at the touch of a finger. On even the most basic e-reader, users can conjure instant definitions without having to hunt through a physical dictionary.

It can be seen that e-readers are superior to printed books. They save space, are environmentally friendly, and provide helpful reading tips and tools that printed books do not. So what good are printed books? Well, they certainly make nice decorations.

   (a)   What is an e-reader?

   (b)   Define e-book. Can e-books exist without a printed edition?

   (c)   Why are e-books better than printed books?

   (d)   How are e-books environment friendly?

   (e)   How does an e-reader make reading a hassle-free experience?

2.   Answer any four from the following questions: [2 × 4 = 8]

   (a)   What made Shaw decide about delivering extempore speeches?

   (b)   What will enrich peoples' lives?

   (c)   Why did Herman stay with her mother?

   (d)   Why was Gandhi against the craze for machinery?

   (e)   Who inspired Kalpana Chawla?

    (f)   Which areas did Kiran Bedi prefer to work in?

    (g)   What is unintelligible to the author of "the Painted Face"?

    (h)   What is the leading source of water pollution today?

3.   Answer any one from the following questions: **[6 × 1 = 6]**

    (a)   'If you look to do something, it is always possible'. Explain.

    (b)   What aspect of modern life does Gardiner satirise in his essay?

4.   Classify any four of the following words as Noun, Verb, Adverb, Adjective: **[1 × 4 = 4]**

Hostile, Inflict, Economically, Suspicion, Obviously, Disrupt, Nationalist, Ambition

5.   Fill in the blanks using appropriate form of verbs from brackets: **[1 × 5 = 5]**

    (a)   Students frequently (make) mistakes of tense usage when they do this exercise.

    (b)   Whenever I (go) to see him, he was out.

    (c)   Most of the class (understand) the results of the experiment yesterday.

    (d)   Production has fallen this year, but it (rise) next year.

    (e)   If you heat this liquid, it (explode).

6.   Fill in the blanks with the most appropriate verbs from the following list: **[1 × 5 = 5]**

can, could, may, might, should, would

    (a)   We …… love our country.

    (b)   I …… not solve this problem.

    (c)   …… you please lend me your pen for a moment?

    (d)   …… we go to watch a film this evening?

    (e)   Probably, he …… lend you money.

7.   Do as directed: **[1 × 4 = 4]**

    (a)   hurray i won the match (Punctuate).

    (b)   Your friends are good. (Add a Question Tag).

    (c)   He runs in the evening. (Make Negative).

    (d)   He is …… student. (Use an Article).

8.   Correct errors from the following sentences: **[1 × 4 = 4]**

    (a)   He saw that the clock has stopped.

    (b)   It has been raining for Monday last.

    (c)   Ramesh is my older brother.

    (d)   Danish, I and you have finished work.

9.   Write a paragraph on *any one* of the following: **[6 × 1 = 6]**

    (a)   Role of technical education for women.

    (b)   Problems of learning English as a Second Language

    (c)   Lokpal Bill as a tool for controlling corruption.

10.   Use any *four* of following words into sentences of your own.   **[1 × 4 = 4]**

vary, very, tear, tier, tail, tale, suite, suit, side, site.

11.   Give antonyms of the given words (*any four*):   **[1 × 4 = 4]**

Accept, Genuine, Come, Wealthy, Dull, Release, Win, Lucky.

12.   Give *synonyms* of the given words (*any four*):   **[1 × 4 = 4]**

Guilty, Lazy, Kind, Assert, Arrive, Hold, Simple, Brave.

13.   Write brief notes on any *four* of the following: **[4 × 4 = 16]**

    (a) Physical appearance   (b) Body language   (c) Speaking skills

    (d) Leadership skills   (e) Group discussion   (f) Telephonic Etiquettes.

✱✱✱